SpringerBriefs in Earth System Sciences

More information about this series at http://www.springer.com/series/10032

Roberto A. Violante · Cecilia Laprida
Natalia L. García Chapori

The Argentina Continental Margin

A Potential Paleoclimatic-Paleoceanographic Archive for the Southern Ocean

Springer

Roberto A. Violante
Division of Marine Geology and Geophysics, Department of Oceanography
Argentina Hydrographic Survey
Buenos Aires
Argentina

Cecilia Laprida
National Council of Scientific and Technical Research, Institute of Andean Studies Don Pablo Groeber, Department of Geology
University of Buenos Aires
Buenos Aires
Argentina

Natalia L. García Chapori
National Council of Scientific and Technical Research, Institute of Andean Studies Don Pablo Groeber, Department of Geology
University of Buenos Aires
Buenos Aires
Argentina

ISSN 2191-589X ISSN 2191-5903 (electronic)
SpringerBriefs in Earth System Sciences
ISBN 978-3-319-04195-7 ISBN 978-3-319-04196-4 (eBook)
DOI 10.1007/978-3-319-04196-4

Library of Congress Control Number: 2017932421

Printed on acid-free paper

This Springer imprint is published by Springer Nature
The registered company is Springer International Publishing AG
The registered company address is: Gewerbestrasse 11, 6330 Cham, Switzerland

Contents

Abbreviations

AABW	Antarctic Bottom Water
AAIW	Antarctic Intermediate Water
AAO	Antarctic Oscillation
ACC	Antarctic Circumpolar Current
ACM	Argentina Continental Margin
AM	Active Margin
AMOC	Atlantic Meridional Overturning Circulation
BMC	Brazil–Malvinas Confluence
BP	Before Present (^{14}C age)
CDW	Circumpolar Deep Water
DIC	Oceanic Dissolved Organic Carbon Reservoir
DSDP	Deep Sea Drilling Program
ENSO	El Niño Southern Oscillation
GDGT	Glycerol Dialkyl Glycerol Tetraethers
ITCZ	Intertropical Convergence Zone
LDEO	Lamont–Doherty Earth Observatory
LGM	Last Glacial Maximum
MCM	Mixed (convergent + sheared) Continental Margin
MOC	Meridional Overturning Circulation
Mwd	Meters water-depth
Myr	Millions of years
NADW	North Atlantic Deep Water
NAO	North Atlantic Oscillation
PMS	Passive Margin Slope
SACW	South Atlantic Central Water
SACZ	South Atlantic Convergence Zone
SAMS	South American Monsoon
SAMW	Subantarctic Mode Water
SHWW	Southern Hemisphere Westerlies
SSS	Sea-Surface Salinity

SST	Sea-Surface Temperature
STSF	Subtropical Shelf Front
TEX86	TetraEther Index of Tetraethers
TMS	Transcurrent and Transpressive Margin Slope
T-SCM	Transcurrent (Sheared) Continental Margin
TW	Tropical Water
VRCM	Volcanic Rifted Continental Margin
WSA	Western South Atlantic Ocean

List of Figures

List of Tables

Abstract

The Argentina Continental Margin, located in the Southwestern Atlantic Ocean, is inserted in a key region of the World Ocean due to its significance in the global oceanographic–climatic interaction. It is the only place where Antarctic- and Equator-sourced water-masses interact in mid-latitudes with a net transport of meridional heat between the Southern Pole and the Equator. On the other hand, the geotectonic history of the region imprints it with significant geological characteristics. As a result, the climatically, oceanographically and physically driven sedimentary processes occurred in the region have originated particular and almost unique morphosedimentary features, which constitute complete records of the processes involved in its evolution. Those features contain different kinds of proxies, tracers and records (biological, geochemical, sedimentological, morphological and structural) which provide valuable quantitative and qualitative evidences for detailed paleoceanographic, paleoclimatic and paleoenvironmental reconstructions. Therefore, the Argentine margin potentially behaves as a complete archive for understanding most of the unique oceanographic and climatic characteristics that occur in the region and impact int he rest of the Southern Hemisphere.

Chapter 1
Introduction

Abstract This chapter introduces to the subjects that will be discussed throughout this book, from the concepts and significance of continental margins to the importance of the Argentine margin in the framework of the global climatic–oceanographic coupling.

Keywords Continental margins · Continent–ocean interaction · Paleoreconstructions · Science and applied science · Hazards · Living and non-living resources

Continental margins are primary and exceptional features of the Earth surface as they represent the transition zones between "pure" continents and oceans, as both continental and oceanic processes have interacted there in relatively recent times of the Earth history, leaving records not only of those processes, but also contributing to the preservation of many features, sediments and rocks representing most of the different earlier stages of the planet evolution. The study of those records constitutes one of the more relevant and valuable sources of information applied for the assessment of climate change, through paleoclimatic, paleoceanographic and paleoenvironmental reconstructions. On the other hand, continental margins store many very important natural resources, both living (fishing and benthic marine faunas) and non-living (hydrocarbons, gas hydrates, minerals, etc.), which in present times are considered to be of a great economic value and possibly one of the most important resources for ensuring life on Earth in the future. Also, natural hazards originated in the marine realm (submarine earthquakes, tsunamis, sea-level rise, floodings of coastal areas, submarine soil instabilities and slope failures with its implications in submarine man-made structures such as drilling platforms, pipes, navigation routes, etc.), as well as pollution, over-exploitation of living and non-living resources, destructive fishing by bottom trawling, uncontrolled urban invasion of seashores, and other man-made impacts, constitute some of the most worrisome problems that society is facing nowadays.

R.A. Violante et al., *The Argentina Continental Margin*,
SpringerBriefs in Earth System Sciences, DOI 10.1007/978-3-319-04196-4_1

Therefore, the study of continental margins is a relevant target in the present-day science objectives. In this context, the Argentina Continental Margin is one of the largest margins in the world, located in an almost unique oceanographic setting containing highly valuable information regarding the evolution of the South Atlantic and also the evidences of significant paleoclimatic and paleoceanographic changes. On the other hand, the region is prone to a number of natural and anthropogenic hazards. However, its study is still uneven, and therefore it becomes necessary to compile the available information with the aim of encouraging further —and absolutely needed—scientific studies in the region. This is the objective of the present book, written as a result of nearly 40 years of work in the Argentine margin done by a multidisciplinary team constituted by marine geologists, geophysicists, sedimentologists, micropaleontologists and oceanographers.

This book is organized in chapters comprising the main subjects related to the present knowledge of the Argentina Continental Margin (ACM). After a brief introduction, in Chap. 2, about the importance of continental margins at a global scale, Chap. 3 contains the evolutionary and geotectonic aspects of the Argentine margin. In Chap. 4 the major morphosedimentary features that shape its relief, as well as the main sedimentary processes modeling it, are described and interpreted. Chapter 5 synthesizes the climatic and oceanographic characteristics of the Southwestern Atlantic that have implications in the margin evolution. Chapter 6 introduces to the principles of paleoceanographic reconstructions, focusing on the concepts of biological, geological, geochemical, as well proxies useful for such reconstructions. Chapter 7 deals with the "state of the art" in the paleoceanographic reconstruction in the Argentina Continental Margin, giving an overview of the present knowledge on the matter. Finally, Chap. 8 contains the conclusions and expectation for future research in the region. A profuse bibliographic list is included, in which the key references related to the subjects contained in the book are mentioned.

Chapter 2
Continental Margins in the Global Context

Abstract As highly significant features of major order at the lands–oceans edge, continental margins play an exceptional role in the Earth system. They are relevant for understanding the continental drift and the birth of oceans, the endogenous and exogenous processes that regulate the planet evolution, the present and past climatic and oceanographic changes, and the carbon cycle and biogeochemistry of the Earth. On this basis, the importance of those features at a global scale is discussed. This chapter also provides the basic definitions needed for understanding the concept of continental margins.

Keywords Continental margins · Continents-oceans boundary · Source-to-Sink · Sedimentary cycle · Sea-floor spreading · Ocean circulation

Continental margins are complex physiographic, geological and biological environments in the transition zone between continents and oceans. The concept of continental margin is closely related to the definition of the boundary between continental and oceanic realms. Although the determination of such a boundary seems an easy matter, complexity soon arises. The shoreline (or "edge of the land," Bascom 1980) is the line of demarcation between the water and the exposed beach (Komar 1976). However, this line simply represents an "instantaneous" position since it changes at a very short—seconds—(waves), daily (tides), yearly (seasonal) and decades (short climatic cycles) scales if the human view point is considered, but also at a geological scale encompassing thousands or millions of years (different orders of sea-level fluctuations, isostasy and tectonism).

Under these considerations it becomes evident that at a large timescale—as is usually the time-frame involved in paleoenvironmental/paleoclimatic reconstructions—a transition zone exists between continents and oceans where the sea–land border has shifted, and the soil and subsoil at each side of the border can share characteristics of both of them, as a consequence of several major facts. (1) Flooding by marine waters of inland regions close to the seashore in recent times of the Earth history (Holocene sea-level transgression) with consequent draping of older continental features by marine sediments, so giving origin to the

R.A. Violante et al., *The Argentina Continental Margin*,
SpringerBriefs in Earth System Sciences, DOI 10.1007/978-3-319-04196-4_2

presently emerged "coastal plains." (2) Exposure of the present seafloor to subaerial conditions during the pre-Holocene sea-level retreat, with consequent development of continental conditions (and hence deposition of terrigenous sediments transported by fluvial, eolian, glacial or any other continental process, and even soils' development) on the present continental shelf. (3) Delivery of terrigenous sediments offshore by marine-related processes, particularly during sea-level retreats, toward deeper, never-exposed sea bottoms of the continental slope and beyond. These considerations have already been explained in more detail by Duxbury (1971) among others. (4) Influence of tectonic and/or isostatic factors that added complexity to the changing position of the shoreline; in this sense, some authors (e.g., Kennett 1982) define continental margins as exclusively related to endogenous factors associated to the Earth deep structure, by saying that they mark the transition between oceanic (dense) and continental (light) crusts (zone of deepening of the Mohorovicic discontinuity) balanced by isostasy.

Therefore, an extremely complex transitional region develops between the "pure" continental and ocean realms, and this region is the "continental margin" (Fig. 2.1), which comprises coastal plains, shelf, slope and rise (Kennett 1982) (Fig. 2.2). Continental margins are considered second-order morphological features

Fig. 2.1 Global map with details of the ocean floor features. The region of the Argentina Continental Margin is indicated by the red square. Observe the "lighter" colors around the continents, indicating the shallowest environments (shelf and part of the slope) in the continental margins. Compare the extension of the Argentine margin respect to others continental margins

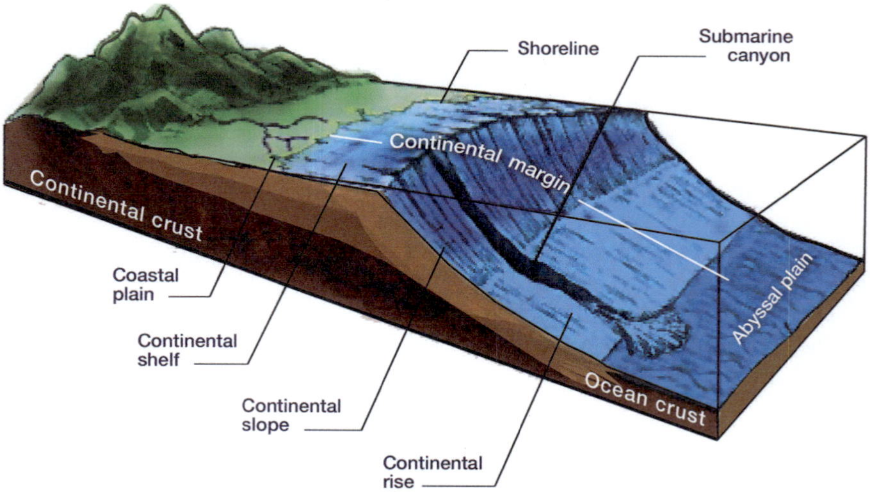

Fig. 2.2 Schematic section of a typical passive margin with its component environments

of the Earth's crust, below first-order features represented by continents and oceans (Heezen and Menard 1966). Continental margins comprise a significant percentage of the world ocean, although there are discrepancies among different authors (e.g., 11% according to COMARGE 2010 and Menot et al. 2010; 15% according to DeMaster 2002; 20% according to Hartnett et al. 1998). Despite these differences, continental margins are of outstanding significance in global sedimentary, biological and geochemical processes, as they store 90% of the total amount of sediments (mostly terrigenous) produced on the Earth surface (McCave 2002) and account for the 25% of the total ocean primary production and more than 90% of all organic carbon burial (Hartnett et al. 1998), and 32% of the biogenic silica deposition in the world ocean (DeMaster 2002).

The term "margin" (sometimes associated to the adjective "continental") is being used since as early as the last part of the 19th century closely related to the evolution of the concept of continental drift and seafloor spreading. Already in 1885, Suess described different types of margins, and these ideas were later incorporated into the Wegener's continental drift theory (in Bond and Kominz 1988). Authors like Johnson (1919), Shepard (1948) and Bourcart (1952) were pioneers in setting the present concepts of continental margins (in Seibold and Berger 1982). Mitchell and Reading (1969) first introduced the term "passive margin" to refer to the earlier concept of "Atlantic-type" margin that had been previously applied to different regions of the world other than the eastern America's region. Since then, the concepts regarding continental margins rapidly evolved into its present application considering the complex set of endogenous and exogenous processes involved in its construction and development.

As the "links" between continents and oceans, either considering endogenous or exogenous processes (or both combined), continental margins are important components in the Earth history as they are relevant in the Source-to-Sink system (Margins 2000) and represent a dynamic link among lithosphere, hydrosphere, atmosphere and biosphere (Nittrouer et al. 2007); hence, they contain the records of many large-, medium-, and low-scale processes involved in the planet evolution, such as

- The continental drift and the birth of oceans
- The geological history of Earth
- The climatic and oceanographic changes that occurred during a large part of the planet evolution (e.g., Wefer et al. 2002)
- The global carbon cycle (Wollast 2002)
- The biogeochemistry, carbon and nutrient fluxes in marine ecosystems (Liu et al. 2010).

Furthermore, continental margins behave as multiple "filters" for sediments being transported from the continent to the sea along the entire sedimentary cycle (Source-to-Sink system). Curray (1975) clearly illustrates this fact (Fig. 2.3) when states that the different environments of the margins (coast, shelf, slope, submarine canyons, rise) are individual features that partially retain terrigenous sediments supplied by rivers, wind, glaciers, volcanic eruptions, coastal and deep currents, etc. Those sediments travel across different environments in a constant search of its final place of deposition. None of the environments is a perfect trap, as they retain only part of the sediments, whereas other materials bypass their boundaries and are transported toward the next (deeper) environment, where again they can be trapped or not, and so on. Recycling of previously deposited sediments as well as locally produced chemical and biogenic materials add new components to the entire sedimentary cycle. The end of the cycle is in the deepest basins where sediments are definitively settled. After that, they suffer several transformations (diagenesis, compaction, consolidation, etc.) and finally they are transformed in sedimentites before entering into the rocks cycle.

Exogenous processes dominating continental margins are mainly associated to the circulation of marine waters, which intervene in the seafloor patterns of sediment transport and distribution, morphosedimentary features, marine biology and geochemistry (Kennett 1982). Marine water circulation depends originally on the solar radiation hitting the planet surface and influencing atmospheric circulation, which drives the wind system that in turn affects the sea surface by "pushing" the water upper levels (e.g., Neumann and Pierson 1966). As these "original" currents travel through different climatic zones and regions of changing water temperatures—and hence the capacity to dissolve salts changes, as well as other physical–chemical properties—different parts of the water masses begun to "float" or "sink" according to density differences. More complexity is incorporated by different degrees of precipitation and evaporation. In this way a distinct vertical stratification develops in the water masses. Sinking waters are then affected by the seafloor morphology when they touch the bottom and can therefore circulate in directions different to the surface waters, particulary if the Coriolis effect is also considered. As a result, thermohaline

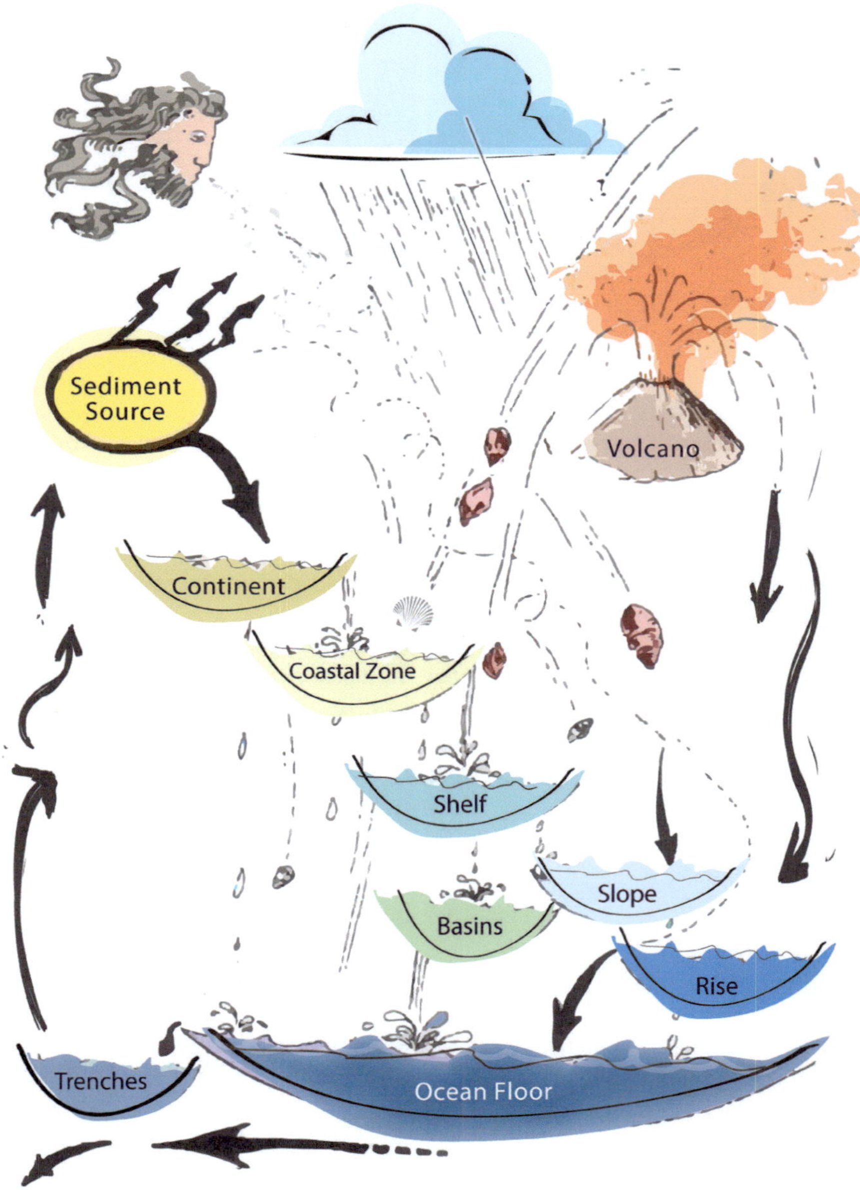

Fig. 2.3 Schematic diagram of the Source-to-Sink system (modified after Curray 1975)

circulation is organized, which basically forms around the world the "Ocean Conveyor Belt" or, as it is nowadays preferred, "Ocean Meridional Overturning Circulation" (e.g., Schmittner et al. 2007) (Fig. 2.4), in which each level of the superposed water masses may run in different directions and at different velocities.

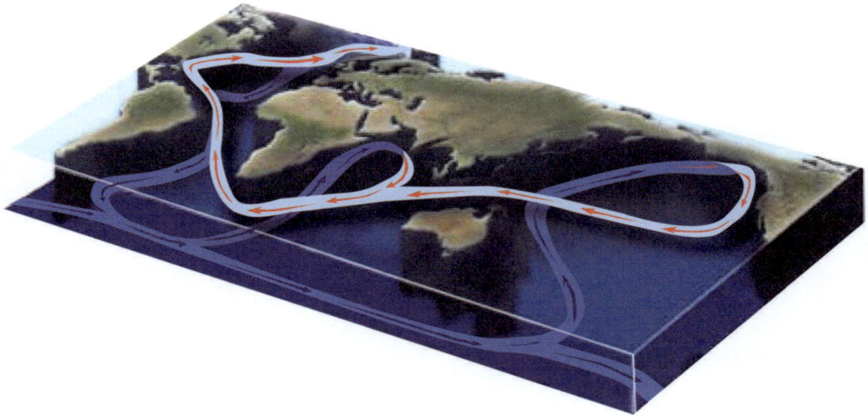

Fig. 2.4 Basic scheme of the main branches of the Ocean Meridional Overturning Circulation

"Local" (regional) factors intervene and create complexity in the entire system, in which high-energy interfaces between adjacent water masses develop and influence the bottom sediments, impacting the seafloor at the edge of continental margins with complex erosive and depositional processes.

In short, it is evident that continental margins are highly complex features that combine continental and marine influence, in which some of the most significant processes affecting the Earth surface occur. In this context, the Argentina Continental Margin occupies one of the most important sectors of the world's margins (Fig. 2.1) as a result of its particular oceanographic setting, as described in the following chapters.

References

Bascom W (1980) Waves and beaches. Anchor Press, 366 pp

Bond GC, Kominz MA (1988) Evolution of thought on passive continental margins from the origin of geosynclinal theory ($\sim$1860) to the present. Geol Soc Am Bull 100:1909–1933

Bourcart J (1952) Océanographie. Introduction á la Géologie. CDU. L´information géographique 16(4):163

Curray JR (1975) Marine sediments, geosynclines and orogeny. In: Fischer A, Judson S (eds) Petroleoum and global tectonics. Princeton Press University, New Jersey, pp 122–157

DeMaster DJ (2002) The accumulation and cycling of biogenic silica in the Southern Ocean: revisiting the marine silica budget. Deep-Sea Res II 49:3155–3167

Duxbury AC (1971) The earth and its oceans. Addison Wesley Publ. Co., 381 p

Hartnett HE, Keil RG, Hedges JI, Devol AH (1998) Influence of oxygen exposure time on organic carbon preservation in continental margin sedimentation. Nature 391:572–574

Heezen BC, Menard HW (1966) Topography of the deep-sea floor. In: Hill MN (ed) The sea. Intersciences Publ, 2nd Edn. J Wiley and Sons: 233–280

Johnson DW (1919) Shore processes and shoreline development. Wiley, New York

Kennett J (1982) Marine geology. Prentice Hall, 813 pp

Komar PD (1976) Beach processes and sedimentation. Prentice Hall Inc., New Jersey, p 429
Liu KK, Atkinson L, Quiñones R, Talaue McManus L (2010) Biogeochemistry of continental Margins in a global context. In: Liu KK, Atkinson L, Quiñones R, Talaue McManus L (eds) Carbon and nutrient fluxes in continental margins: a global synthesis. The IGBP Series, Springer, pp 3–24
Margins (2000) Source-to-Sink studies. Margins–NSF Program, Newsletter 5. http://www.nsf-margins.org
McCave IN (2002). Sedimentary settings on continental margins: an overview. In: Wefer G, Billet D, Hebeln D, Jorgensen BB, Schluter M, van Weering TCE (eds) Ocean margin systems, pp 1–14
Menot, L, Sibuet M, Carney RS, Levin L, Rowe GT, Billet DSM, Poore G, Kitazato H, Vanreusel A, Geléron J, Lavrado HP, Sellanes J, Ingole B, Krylova E (2010) New perceptions of continental margins biodiversity. In: McIntyre A (ed) Life in the world ocean: diversity, distribution and abundance. Willey Blackwell, 384 pp
Mitchell AH, Reading HG (1969) Continental margins, geosynclines and ocean floor spreading. J Geol 77:629–646
Neumann G, Pierson W Jr (1966) Principles of physical oceanography. Prentice-Hall Inc., NJ 545 pp
Nittrouer CA, Austin Jr JA, Field ME, Kravits JH, Syvitski JPM, Wiberg P (2007) Writing the Roseta Stone: insights into continental-margin sedimentary processes and strata. In: Nittrouer CA, Austin Jr JA, Field ME, Kravits JH, Syvitski JPM, Wiberg P (eds) Continental margin sedimentation, from sediment transport to sequence stratigraphy. International association of sedimentologists, Sp. Publ. N° 37, Blackwell Publishing, pp 1–48
Schmittner A, Chiang JCH, Hemming SR (2007) Introduction: the Ocean's meridional overturning circulation. In Schmittner A, Chiang JCH, Hemming SR (eds.) Ocean circulation: mechanisms and impacts—past and future changes of meridional overturning. American Geophysical Union, Washington, D.C., pp 1–4. doi:10.1029/173GM02
Seibold E, Berger WH (1982) The sea floor. an introduction to marine geology. Springer, Berlin–Heidelberg–New York 288 p
Shepard FP (1948) Submarine geology. Harper & Row, New York
Wefer G, Billet D, Hebeln D, Jorgensen BB, Schluter M, van Weering TCE (2002) Ocean margin systems. Springer, Berling–Heidelberg 495 pp
Wollast R (2002) Continental margins: a review of geochemical settings. In: Wefer G, Billet D, Hebeln D, Jorgensen BB, Schluter M, van Weering TJCE (eds) Ocean margin systems 15–31

Chapter 3
The Argentina Continental Margin: Location and Significance

Abstract The Argentina Continental Margin (ACM) is one of the largest margins on the Earth. It is located in a key region of the World Ocean that is highly significant in the planet's oceanographic-climatic system. The evolution of the Argentine margin is explained in terms of the combination and interaction among geotectonic, oceanographic and climatic factors. Because of that evolution, four types of margins develop in the region: a passive-volcanic rifted, a transcurrent-sheared, a mixed and an active margin. These aspects determine particular tectonic and stratigraphic characteristics for each of them.

Keywords Argentina Continental Margin · Water–masses exchange · South Atlantic Ocean · Geotectonic evolution · Passive and active margins · Sedimentary basins

The ACM, one of the largest margins worldwide, is a complex setting in the Western South Atlantic Ocean (WSA) (Fig. 3.1). It constitutes a key region in the global oceanographic-climatic system (Ej.: Hastenrath 1982; Díaz and Markgraff 1992; Berger and Wefer 1996; Wefer et al. 1996, 2004; Markgraff 2001; Mata et al. 2001; Bryden and Imawaki 2001; Talley 2003; Franke et al. 2007; Carter and Cortese 2009) as is the only place in the southern ocean with a net water–masses exchange between south polar and equatorial regions (Fig. 3.2). Strong Antarctic-sourced currents run northwards along the margin driven by the Coriolis force from 56°S reaching regions up to at least 34°S and even north. At 38°S those currents meet the equatorial-sourced currents that run southward along the Brazilian margin, originating the so-called Brazil–Malvinas Confluence (BMC) and the Subtropical Shelf Front (STSF). The structure of the entire oceanographic system, as it will be described in following chapters, is composed of different superposed water-masses characterized by significant changes in density, temperature, salinity, oxygen content and other physical, biological and chemical properties, so developing a very complex stratified structure (Table 3.1). On the other hand, offshore

R.A. Violante et al., *The Argentina Continental Margin*,
SpringerBriefs in Earth System Sciences, DOI 10.1007/978-3-319-04196-4_3

Fig. 3.1 Location of the Argentina Continental Margin in the context of South America. Note the extension of the margin (*light blue color*) in relation to the extension of Argentina

the ACM, the neighboring province of the Argentine Basin (Fig. 3.1) is known as probably one of the largest containers of deep-sea sediments in the World Ocean (Emery and Uchupi 1984), and in this sense the margin (shelf+slope+rise) is the area through which most of those sediments are transferred (and partially momentaneously or definitively deposited) from coastal and inland areas to the basin.

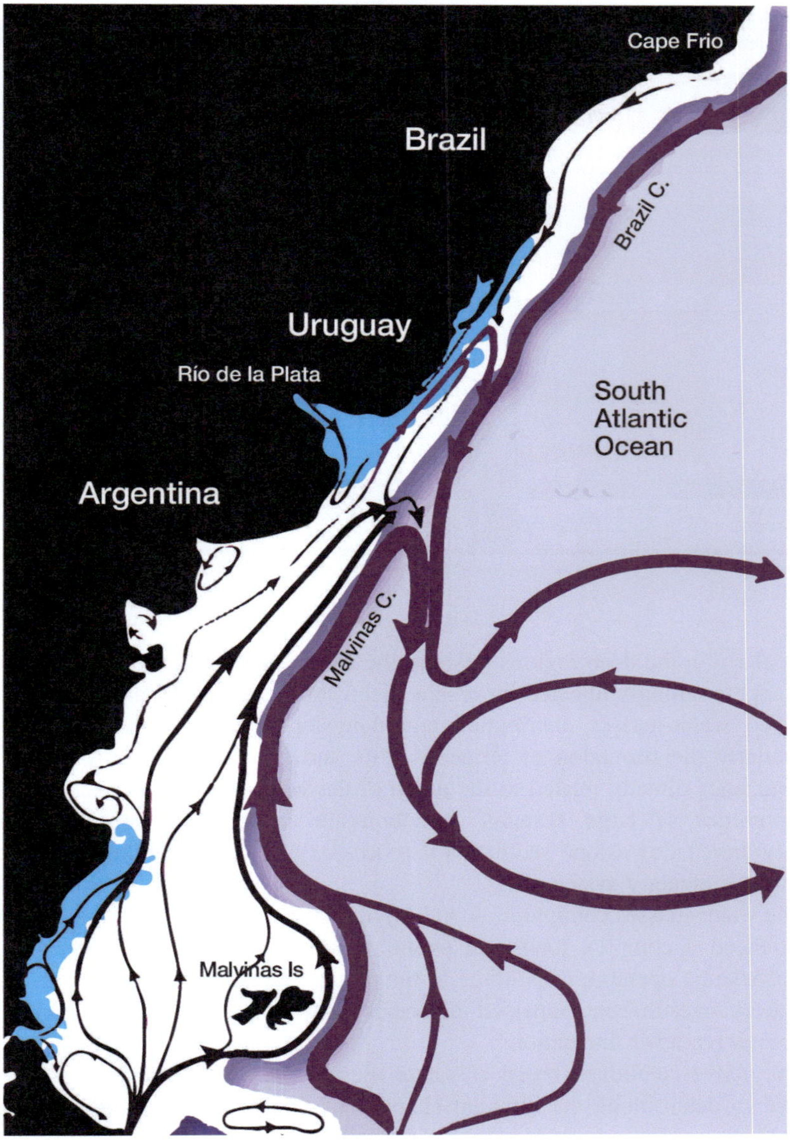

Fig. 3.2 Main surface currents that compound the basic circulation in the Western South Atlantic Ocean (modified after Matano et al. 2010). Note the region (in front of the Río de la Plata) where north-sourced and south-sourced water-masses meet (Brazil–Malvinas Confluence)

The margin morphology encompasses the coastal plains, a wide continental shelf and extended, stepped slope and rise, so constituting as a whole a broad and extensive surface starting in the coast and reaching depths up to ~5000 m (Fig. 3.3). This enormous environment is subjected to a great variety of processes.

Table 3.1 Distribution of water-masses in the Western South Atlantic (modified after Hernández-Molina et al. 2010). For acronyms of the water-masses see text

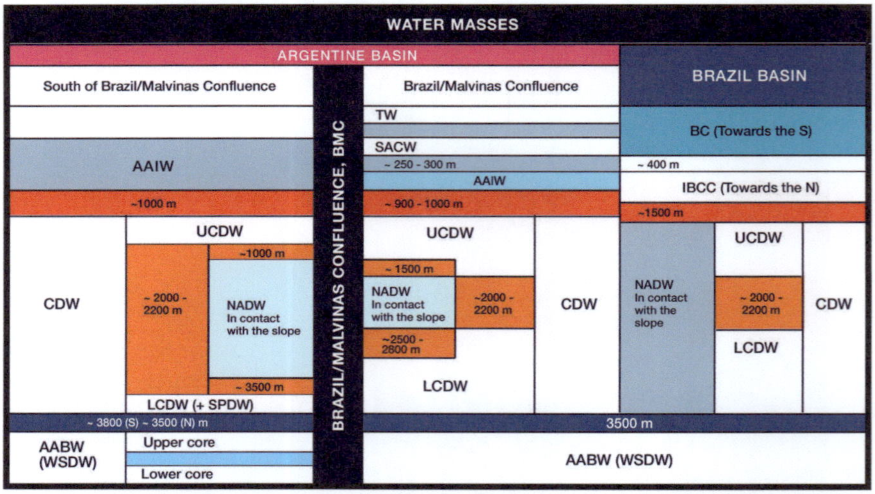

Stratification of the water masses in the Argentine and Brazil Basins, indicating the depth of the interfaces.

Particularly in the deep regions offshore the continental shelf, morphosedimentary features are strongly affected by erosive-depositional processes resulting from the stratified water-masses distribution and alongslope bottom currents circulation, particularly the formation of terraces, drifts and associated features that follow contour lines directly related to the depth of the water-masses interfaces. Besides, the presence of large systems of submarine canyons as well as energetic gravity-driven downslope sedimentary processes add complications to the whole morphosedimentary system.

The water-masses configuration was highly variable through time, as the margin experienced a complex long-term evolution since the Gondwana break-up and South Atlantic opening, responding to the oceanographic evolution and associated climate changes that accompanied the ocean expansion as well as the occurrence of numerous sea-level fluctuations.

The above-mentioned broad characteristics of the margin, which will be considered in detail along the different chapters of this book, make the whole environment a complex system that stores thick sedimentary sequences, whose deposits, faunal content, geochemistry and other characteristics, studied through different proxies and records, must reflect the complex water-masses/sediments interaction and dynamics and therefore a relatively detailed reconstruction of the sedimentary and paleoceanographic processes and resulting features can be performed. Added this to the fact that the ACM occupies a key (and almost unique) region in the global ocean, as stated above, makes the region a undoubtedly highly potential archive for global, hemispheric, regional and local climatic, oceanographic, morphosedimentary, geochemical and biological changes.

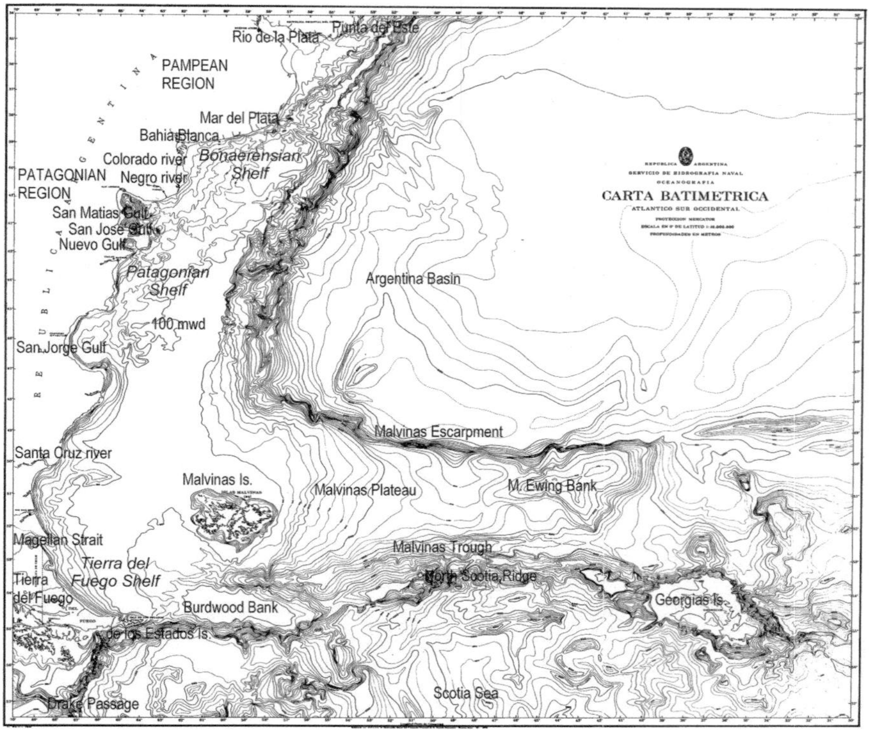

Fig. 3.3 Bathymetric map of the Argentina Continental Margin. *Source* Bathymetric Chart of the Western South Atlantic, Argentina Hydrographic Survey (1976). Geographic references mentioned in the text

3.1 Geotectonic Considerations, Types of Margins and Stratigraphy

The geotectonic framework into which the ACM evolved, conditioned the different stages of development through diverse interactions among climatic, oceanographic and tectonic factors. The present-day margin configuration, types and sedimentary constitution are the result of that combination of factors.

3.1.1 Brief History of the Margin Evolution in the Context of the Southern Hemisphere

The geotectonic evolution of the South Atlantic Ocean is primarily related to the seafloor spreading and separation between South America and Africa plates, in a region of cortical extension associated to the Gondwana break-up. Consequently, the margin structure is highly complex as it is the result of the sum of different continental blocks formed in successive orogenic phases (Ramos 1996; Urien and Zambrano 1996; Hinz et al. 1999; Max et al. 1999; Franke et al. 2007). The initial stages of the South Atlantic evolution occurred in mid-Jurassic times with the occurrence of rift and wrench processes (Ladd 1974; Rabinowitz and LaBrecque 1979; de Witt et al. 1988; Etheridge et al. 1989; Lister et al. 1991; Urien and Zambrano 1996; Ramos 1996, 1999; Turic et al. 1996). Following that, voluminous volcanic effusions took place in Late Jurassic–Early Cretaceous times (Hinz et al. 1999; Franke et al. 2007, 2010). The separation between South America and Africa was completed with the reactivation of the Malvinas–Agulhas fracture in the Aptian (around 115 Myr), when sea-waters invaded former euxinic environments and a proto-Atlantic ocean was installed (Rabinowitz and LaBrecque 1979; Hinz et al. 1999). In the Early Campanian (81 Myr) deep waters already played an important role in shaping the sea-floor as evidenced by complex sedimentary drifts in the Southwestern Atlantic (Hinz et al. 1999). Starting in the Maastrichtian, a global transgressive event took place, which gave later origin to the first Cenozoic marine transgression affecting southeastern South America, with characteristics of a very shallow sea (Malumian 1999). At these times of the Late Cretaceous, a compressive tectonic regime (Andean Orogeny) was initiated in the western margin of the South America Plate (Dalziel et al. 1974; see also Key et al. 2009 as a general reference). Diverse stages of this orogeny occurred until the first epochs of the Cenozoic, being the more relevant fact the counter-clock rotation of the southern extreme of the proto-Andes Cordillera (Rapalini 2007 and references therein cited) that gave origin to the development of the Scotia sea. At the beginning of the Eocene (55–40 Myr), when South America and Antarctica were still connected and the proto-Atlantic ocean was already well developed, sea level was globally high in a climatic context of high temperatures; surface and bottom ocean temperatures at low latitudes were typical of subtropical seas. This high sea-level event reached some sectors of the present Patagonian lands according to micropaleontological evidences reported by Malumian and Nañez (2009). As this ocean progressively evolved into an open sea, thermohaline circulation developed, which in the first stages was mainly controlled by salinity rather than by temperature (Oberhansli and Hsü 1986).

At the end of the Eocene (40–35 Myr) the global cooling of the deep water-masses enabled a significant marine regression and also the displacement of the zones of formation of deep waters from low to high latitudes. At 34 Myr, the first evidences of formation of ice-masses appeared in East Antarctica (Einsele 2000) with a global decreasing in temperatures that averaged 4–5 °C (Zachos et al. 2001). In the Upper Eocene, global tectonic deformations were documented, producing particular effects in western South America, where the Inca tectonic phase related to the beginning of the Andes rise took place.

Starting in the Early Oligocene (around 32–30 Myr, Livermore et al. 2004; Lawver and Gahagan 2003) the expansion of the Scotia Plate to the east enabled the opening of the Drake Passage (together with the opening of the Tasmania Passage that started earlier, around 38 Myr), what led to the isolation of Antarctica and the subsequent initiation in the circulation of the Antarctic Circumpolar Current (Einsele 2000; Zachos et al. 2001). This event set the model of the present thermohaline circulation in the ocean. The Drake Passage reached its complete opening at 29 Myr, at the same time that the Pehuenche diastrophic phase occurred in the Andean region. As a consequence, the heat transference driven by the marine circulation between low latitudes and Antarctica was interrupted, and East Antarctica responded to it with a big cooling and expansion of the ice-masses that induced a significant regional sea-level fall.

At the same time, the occurrence of the first stages of the NW–SE sea-floor spreading and transtensional motions in the Scotia plate enabled it to penetrate beneath the Drake Passage, with the consequent formation of the Scotia Arc. This event lasted for a long time between 30 and 6 Myr (Livermore et al. 2004; Lodolo and Tassone 2010; Dalziel et al. 2013 and references therein cited). In the meantime, in the Late Oligocene (25 Myr) a global warming with sea-level rise and associated marine transgressions occurred; some parts of Patagonia were affected by this transgression, which is recorded as very shallow and limited-extended seas (Malumian and Nañez 2009). Soon after, in the Lower Miocene (23 Myr), a new global cooling with low sea levels conduced to a new glaciation (Zachos et al. 2001).

The Miocene continued, between 17 and 14.5 Myr, with another increasing in temperature, reduction in the Antarctic ice-masses and a third-order sea-level rise at a global scale (Haq et al. 1987; Zachos et al. 2001), that together with a significant regional subsidence (van Andel et al. 1977; Kennett 1982) gave rise to Atlantic marine transgressions represented in the Pampean regions of Argentina by the Paranense transgression, and in Patagonia by the marine facies described by Malumian and Nañez (2009); according to Berggren et al. (in Malumián and Nañez 1996), the climatic optimum of the Neogene occurred in the Middle Miocene between 15.6 and 13.6 Myr. At these times, the North Atlantic Deep Water (NADW) first influenced the Southern Hemisphere (Preu et al. 2012). Following these periods, the ice-masses grew up again affecting the whole Antarctic continent, and the circulation of the Antarctic Bottom Water (AABW) begun to be very active (Einsele 2000). This event gave rise to a new marine regression. Simultaneously, the orogenic Quechua phase (that encompassed the time span between 14 and 10 Myr) produced the final uplifting of the Andes (Patagónica and Principal Cordilleras) when they nearly reached its present configuration (Yrigoyen 1979, 1999), and strong subsidence was evidenced in the Southwestern Atlantic (WSA) (e.g.: Kennett 1982; Aceñolaza 2000; Potter and Szatmari 2009). It enabled the installation of particular geomorphological conditions, as a substantial increasing in sediment supply from the new high-relief mountains in the west toward the sea was favored. At the same time, the WSA experienced the circulation of the Antarctic Intermediate Water (AAIW).

These climatic-oceanographic-tectonic conditions lead to a new decreasing in the oceans temperature by around 4 °C, and the definitive and permanent installation of the Antarctic ice-masses. The cold climate got installed in the Patagonian–Pampean region and the first evidences of glaciers advances in Patagonia were registered at 7 Myr (Rabassa et al. 2005). At around 5–4 Myr the global cooling was interrupted and temperatures got slightly warmer again at 4–3 Myr. During the Late Pliocene, a new Andean diastrophic phase (Diaguita) occurred, being responsible for the final uplift of the central Andes as well as other orographic systems of Argentina (Puna and Pampean Mountain Range), and also the uplift of the Mesopotamian region. Additionally, during the Late Pliocene (3–2.4 Myr) a significant event at a global scale occurred, which was the definitive closing of the Panamá Isthmus. It produced the increasing activity of the surface Gulf Stream that begun to transport warm and saline waters towards the North Atlantic, accompanied by the intensification of the North Atlantic Deep Water (NADW) circulation, the formation of ice-masses in the Northern Hemisphere and the beginning of the Quaternary glaciations. The increasing in temperature gradients influenced the circulation of the NADW (Einsele 2000). Glacial conditions were definitively settled in Antarctica and Patagonia. During the Quaternary, the AABW definitively reactivated and the deep-water circulation reached its present configuration, with an increasing of the AABW during glacial periods and the NADW during interglacial periods (Duplessy et al. 1988; Sarnthein et al. 1994; Flower et al. 2000; Rahmostort 2006; Laprida et al. 2011; Bender et al. 2013; Voigt et al. 2015).

Through all this history, three major processes dominated the regional evolution and were responsible for defining four types of margins in the WSA (Fig. 3.4): (a) the westward motion of the South America Plate; (b) the transcurrent movement relative to the Malvinas–Agulhas fracture; (c) the interplay between South America and Scotia Plates, the Malvinas–Agulhas fracture and the subduction zone in the Sandwich Plate; and (d) Active margin en the South Sandwich trench. Relative movements between plates are indicative of the regional dynamics. Pelayo and Wiens (1989) and Barker et al. (1991) estimated in 7.5 cm/year the relative subduction velocity between South America and Sandwich Plates, with variable seismic activity depending on the decreasing subduction angles from north of 57°S to south of 59°S (Brett 1977; Frankel and McCann 1979; Barker et al. 1991). Particularly, the Sandwich oceanic ridge has been active at least in the last 7 Myr (Ewing et al. 1971), with increasing expansion velocity from 5–6 cm/year to near 9 cm/year (Barker 1972).

3.1.2 Margin Types and Geotectonic Configuration

In response to the geotectonic evolution, four types of margins were configured in the WSA (Ramos 1996; Hinz et al. 1999; Franke et al. 2007, 2010) (Fig. 3.4). Most of the geographic references mentioned in the text are indicated in Fig. 3.3. Otherwise they are included in other figures.

Volcanic rifted continental margin (VRCM in Fig. 3.4): also named "Volcanic Extensional Continental Margin" (Mohriak et al. 2002), corresponds to a typical lower-plate passive margin with rift basins (Ramos 1996) associated to sea-floor spreading. It is characterized by a young and thin crust with pre-rift associations and longitudinal rifts, and was strongly affected by volcanism. This margin extends from eastern Brazil at 30°S (although some authors as Mohriak et al. 2002 and Gladczenko et al. 1997 extend it even northward up to 20°S) to ~48–49°S in southern Patagonia, so that most of the ACM fits into this type. This margin is structured in five segments at least in the Argentine and Uruguayan sectors. Segments named I, II, III and IV

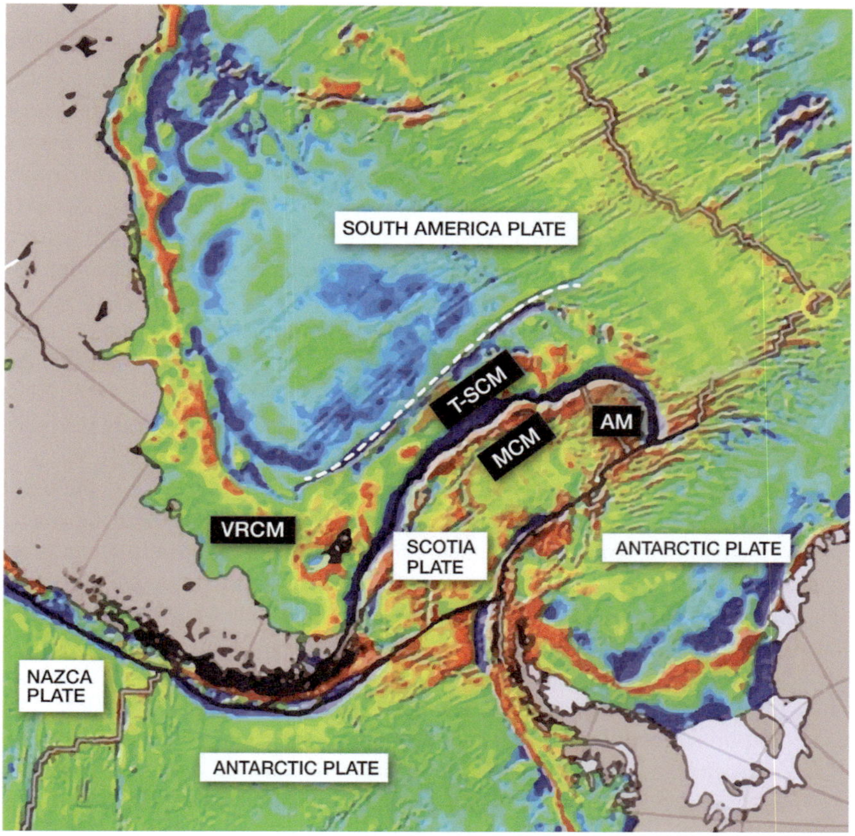

Fig. 3.4 Geotectonic configuration of the Western South Atlantic and types of margins in the Argentina Continental Margin (modified from Ghidella et al. 2007. Base map from satellite gravimetric data from Sandwell and Smith 1997; McAdoo and Laxon 1997). The *dotted white line* indicates the location of the western extension of the Malvinas–Agulhas fracture zone. *VRCM* Volcanic rifted continental margin. *T-SCM* Transcurrent (sheared) continental margin. *MCM* Mixed (convergent + sheared) continental margin. *AM* Active margin

(from south to north, all of them belonging to the Argentina Continental Margin, Franke et al. 2007) are delimited at their southern boundaries by Transfer Fracture Zones (Malvinas, Colorado, Ventana and Salado respectively). The Colorado Transfer Zone, identified by magnetic (Ghidella et al. 1995) and seismic (Franke et al. 2007) data, separates the "Bonaerensian" and the "Patagonian" domains (Max et al. 1999; Ramos et al. 2004; Cavallotto et al. 2011). A fifth segment was identified in the Uruguay Continental Margin by Soto et al. (2011), who found a new transfer zone (named Río de la Plata) in the upper section of Segment IV, so redefining it and introducing the so-called Segment V north of that transfer zone. To the north, several fracture zones and transverse basins characterize the continuation of the same type of margin in the Southern Brazilian Margin (e.g.,: Bassetto et al. 2000).

The main characteristics of this margin in the Argentine sector can be synthesized after Hinz (1981), Mutter et al. (1982, 1988), Ramos (1996), Gladczenko et al. (1997), Hinz et al. (1999) and Franke et al. (2007), among others, as represented by: (a) a basement of continental crust structured both parallel and perpendicular to the margin by extensional (listric) faults, parallel systems generating hemigrabens; (b) associations of perpendicular to slightly oblique fractures characterizing aborted rifts or aulacogens (de Witt 1977); (c) a basement of oceanic and continental crust affected by Transfer Fracture Zones perpendicular or oblique to the margin; (d) external peripheric highs (or domes) corresponding to basement´s highs; (e) a transition zone between the continental and the oceanic crusts characterized by a set of seaward dipping, high amplitude, upward convex, thick seismic reflectors associated to ancient volcanic effusions and evidenced by the so-called "G" magnetic anomaly (Light et al. 1993; Ramos 1996; Parker et al. 1996; Gladczenko et al. 1997; Hinz et al. 1999; Neben et al. 2005; Franke et al. 2007); (f) main tectonic structures are continuous and related to inland structures; and (g), extensive, broad epicontinental shelves.

Transcurrent (sheared) continental margin (T-SCM in Fig. 3.4): is the part of the margin associated to the Malvinas–Agulhas fracture, a complex transfer fracture zone represented by a strike-displacement fault, along which the southern margin of the South America Plate was displaced to the west during the continental separation. The major feature characterizing this margin is the Malvinas Escarpment located at 48–49°S, which constitutes the northern boundary of the Malvinas Plateau (that includes the Malvinas Islands and the M. Ewing Bank), a complex feature resulted from the interaction between South America and Scotia Plates that shares morpho-structural characteristics of the four types of margins.

Mixed (convergent + sheared) and active continental margin (MCM in Fig. 3.4): corresponds to the Scotia Arc, a complex tectonic element that comprises fragments of ancient blocks of continental crust and former volcanic arcs moved from the southern extreme of the Andes Cordillera through displacement faults that enable to join the extreme of the South America continent with the continental blocks of the San Martín Land (or Palmer Land) in Antarctic Peninsula (Dalziel and Elliot 1971; de Witt 1977; Barker and Dalziel 1983). The northern part of the Arc is the North Scotia Ridge, a feature that represents the partially submerged extension of the Andean cordillera, composed of continental blocks of deeply deformed sedimentary, leptometamorphic and igneous rocks (Dalziel 1982); it extends through de los Estados Island, Burdwood Bank and Georgia Islands to the east and ends at the South Sandwich Volcanic Arc, this one representing an active volcanic feature associated to a microplate in a subduction zone at the boundary between oceanic plates, and therefore considered by some authors as a different type of margin (active, intra-oceanic arc, AM in Fig. 3.4) (Pelayo and Wiens 1989; Leat et al. 2003). The Malvinas Trough represents the boundary between the transcurrent and the mixed margins, i.e. the boundary between South America and Scotia Plates.

Configuration of the ACM: The configuration and characteristics of the ACM are regionally variable in response to the evolutionary history. Ramos (1996) considers that the sector of the passive margin north of 43°S has an ancient basement and thicker continental crust, its evolution having been controlled by previous cortical discontinuities that developed extensional systems transverse to the margin with little basaltic magmatism. On the other hand, the sector of the passive margin between 43 and 49°S has a younger and less-thick crust with pre- and early-rift pre-tectonic associations characterized by acid volcanism and extensions that provoked longitudinal rifts. Instead, the mixed and active margins in the south experienced a more complex history associated to the eastward motion of the oceanic Scotia Plate being inserted as a wedge between South America and Antarctic Plates.

On the basis of all the above-mentioned evolutive factors, the ACM presents two major features (Emery and Uchupi 1984; Turic et al. 1996), which are particularly better defined in the shelf region. North of 55°S (comprising the volcanic rifted—VCRM—and the transcurrent/sheared—T-SCM—margins) the shelf was affected by downwarping processes conditioned by the isostatic equilibrium and sedimentary overloading, which conduced to the formation of very thick sedimentary sequences in which most of the sediment supply was enhanced after the Miocene by the rising of the Andes Cordillera. On the other hand, south of 55°S (mixed and active margins—MCM and AM) complex tectonic processes led to the development of a thin Cenozoic sedimentary coverage that was then deeply affected by marine erosion. The basins that constitute the margin's substratum show regional structural features of extensional systems oriented both along and transverse to the margin. At the land–sea boundary, the shoreline configuration is inherited from the geological and structural features of the basins. The sea-floor conserves the same features although obliterated by the Cenozoic sedimentary coverage.

The evolutionary, tectonic, structural and morphosedimentary characteristics of the ACM enable to classify it, according to Heezen and Menard (1966), as a second-order feature including third-order categories I, II and III. They contain fourth-order features like continental shelf, shallow epicontinental sea, continental slope and continental rise in the passive sector of the margin, as well as plateaus, escarpments, trenches and islands in the transcurrent, mixed and active margins. All of them on its turn are composed of fifth-order features like terraces, scarps, steps and submarine canyons.

3.1.3 Stratigraphic Evidences of the Margin Evolution

The geotectonic history of the WSA and southern South America originated the formation of marginal, transverse-to-the-margin sedimentary basins (Fig. 3.5). They hold post-Cretaceous sequences above a pre-Cretaceous basement which is also present in the interbasin areas (Zambrano and Urien 1974; Ramos and Turic 1996; Ramos 1999 for the Argentinian margin; Meisling et al. 2001 for the Brazilian margin). The post-Neogene sedimentary sequences have the peculiarity that they are not restricted to the post-Cretaceous basins but overpass their limits and extend homogeneously on most of the Argentina continental and marine territory. On the basis of a numerous of available offshore drillings performed for hydrocarbon exploration, two sedimentary sequences are recognized in the

Neogene parts of the basins: the lower sequence (aged Miocene to Lower Pliocene) is integrated by marine deposits formed during the Miocene transgression—which covered most of South America, Aceñolaza (2000)—as well as by regionally extended Lower Pliocene continental (mostly fluvial) deposits. The upper sequence (aged Pliocene to Pleistocene) is made up of loessic (eolian-lacustrine) deposits interdigitated with marine sediments that resulted from three glacieustatic (transgressive-regressive) events (Violante and Parker 1993, 1999; Parker et al. 1999, 2008). The correspondent transgressions affected only the shelf and coastal regions as a consequence of the continental uplift that followed the Andes Cordillera formation and precluded deeper inland penetration of marine waters. The last Plio-Pleistocene sea-level (preceding the post-Last Glacial Maximum—LGM—transgression) corresponds to isotopic stage 5e (Isla et al. 2000; Violante 2003; Parker et al. 2008). The two sequences are separated to each other by a discontinuity (seismic horizon "b" defined by Ewing and Lonardi 1971), which was recognized at 140 m water depth in the subsurface of the shelf offshore the Rio de la Plata mouth. This horizon extends to the south reaching $\sim$43°S. Consequently, the substratum of the post-LGM deposits is represented, north of 43°S, by nearly complete Plio-Pleistocene marine sequences, whereas south of 43°S is mainly represented by Late Tertiary pre-glacial continental and marine sequences with scattered and incomplete patches of Plio-Pleistocene deposits. The discontinuous and reduced Quaternary deposits in the southern region can be attributed to lack of space for deposition due to the post-glacial isostatic rebound, which was more significant there, closer to the glaciated areas of Patagonia (Codignotto et al. 1993; Rostami et al. 2000).

A particular feature of the margin is the so-call Argentine Basin (Fig. 3.5), an along-slope very large sedimentary basin containing an exceptionally thick sediments accumulation, whose tectonic evolution occurred during and following continental break-up (Lonardi and Ewing 1971; Tankard et al. 1995; Thomson 1998; Hinz et al. 1999; Macdonald et al. 2003; Franke et al. 2006, 2007).

The application of the seismic-stratigraphic concepts with consequent recognition of seismic reflectors representing regional unconformities that separate seismic sequences associated to depositional sedimentary bodies, was a major advance in the knowledge of the sedimentary processes and the timing of the different events involved in the evolution and construction of the Argentine margin. Table 3.2 summarizes the progression in the identification and interpretation of seismic reflectors and sequences, compiling all the bibliographic references and advancing in the seismic correlations. As it can be seen there, a numerous of seismic reflectors were recognized, each of them being the consequence of interacting tectonic, oceanographic and climatic changes that left relevant records in the entire stratigraphic sequence. The more striking changes that affected the Argentine margin configuration were the modifications in the thermohaline circulation produced by the onset of the Antarctic-sourced bottom waters in the Eocene–Oligocene transition, the later diverse reactivations in the northward-flowing alongslope currents, and the growing up and development of the contouritic depositional systems that constitute the main processes in building and shaping the margin.

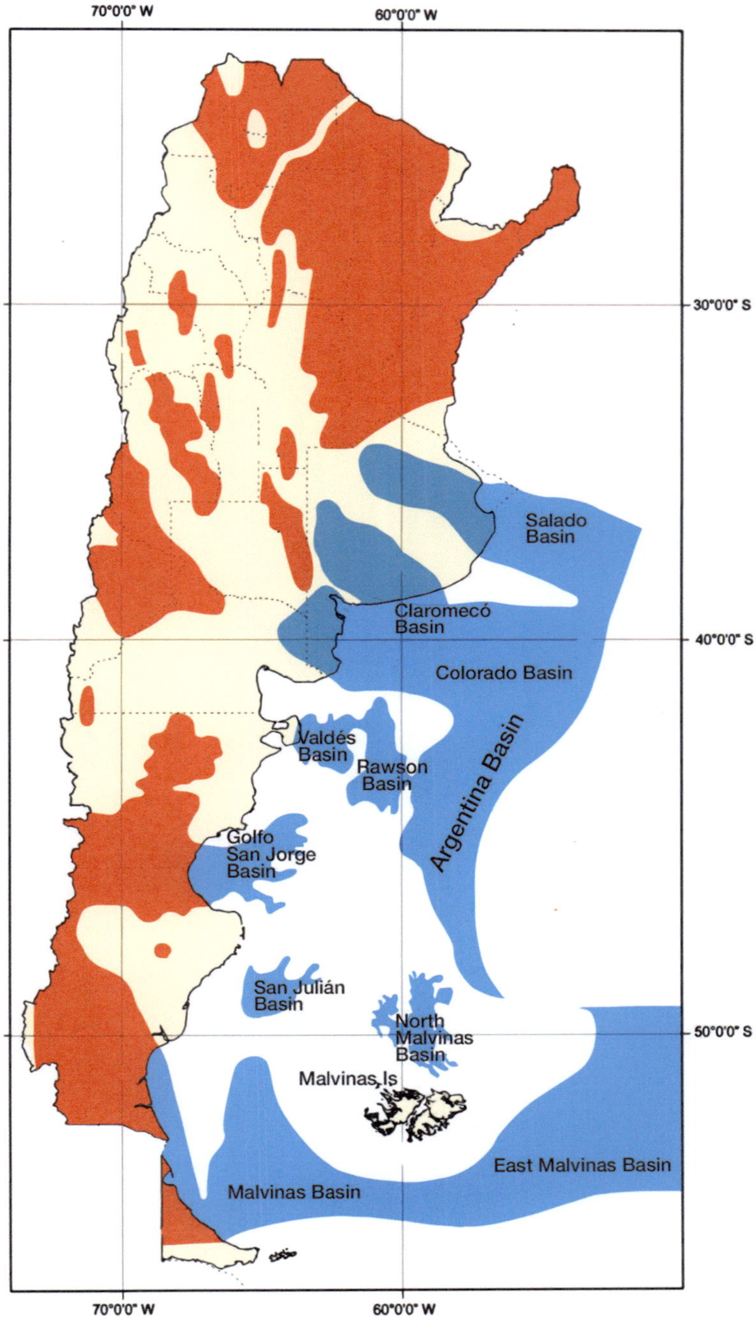

Fig. 3.5 Sedimentary basins in Argentina. In *blue*, the offshore basins

Table 3.2 Seismic reflectors in the Argentina Continental Margin, which define the major seismic-stratigraphic units

Ewing and Lonardi (1971)	Urien and Zambrano (1996)	Hinz et al. (1999)	Franke et al. (2007)	Parker et al. (1999, 2005, 2008)	Schumann (2002)
b				N	
H1				L	
H2 Eocene					ARG 9 (5 Myr)
H3 Eocene		AR5 Mid-Miocene (~15 Myr)			(~12 Myr)
	ARI Mid Cenozoic unconformity (Oligocene paleo-platform?)	AR4 Upper Eocene			ARG 6 Eocene-Oligocene (~30 Myr)
H4 Top of Cretaceous	ARII Cretaceous-Paleocene boundary, Maastrichtian transgression	AR3 Upper Cretaceous-PreMaastrichtian (~81 Myr)	Pedro Luro equivalent (PLe)-Cretaceous-Tertiary boundary		ARG 7 Cretaceous-Tertiary boundary
		AR2 upper Aptian			ARG 4
		AR1 Hauterivian (~125 Myr)	BU Break-up unconformity		

Violante et al. (2010a, b)	Gruetzner et al. (2011, 2012)	Hernández Molina et al. (2009)	Preu et al. (2012)	Significance	Age
N				Base of the Quaternary glacieustatic transgressive-regressive	2.4 Ma
H1-L	VF4 (3 Myr)	H1	H1 (= VF3 of Gruetzner et al. (2011)	Increases AABW, CDW and NADW activity	Lower-mid Pliocene boundary

(continued)

Table 3.2 (continued)

Violante et al. (2010a, b)	Gruetzner et al. (2011, 2012)	Hernández Molina et al. (2009)	Preu et al. (2012)	Significance	Age
H2	VF3 (6 Myr) = AR7	H2	H2 (= VF1 of Gruetzner et al. (2011)		Miocene-Pliocene boundary
	VF2 (9 Myr) VF1 (12 Myr)				Late Miocene
	AR6 (14 Myr)			Increment in deep-water circulation and strengthening of the Antarctic Circumpolar Current (ACC). Circulation of the AABW begun to be very active and the contouritic drift in the slope begun to reach their major development. Also, transverse-to- the-slope processes associated to increasing terrigenous sediment input to the margin begun to be significant, as a result of the tectonic uplifting of the Andes Cordillera and consequent increase in the gradient of surface continental runoff	Represents the mid-Miocene transition (Post MMC0)
AR5	AR5 (17 Myr)	AR5	AR5	Expansion of the Antarctica ice-sheet with significant cooling of Antarctic Peninsula	Early-mid Miocene. Represents the pre-MMCO (Mid-Miocene Climatic Optimum). 16 Myr
R*					Oligocene-Miocene boundary

(continued)

Table 3.2 (continued)

Violante et al. (2010a, b)	Gruetzner et al. (2011, 2012)	Hernández Molina et al. (2009)	Preu et al. (2012)	Significance	Age
AR4	AR4 (34 Myr)	AR4	AR4	Big expansion of East Antarctica ice-sheet towards the Atlantic. Consequent deep modification of thermohaline circulation, onset of Antarctic bottom current activity (particularly AABW). Drastic modification in the sedimentary depositional regime in the Argentine basin, changing from hemipelagic-dominated in the Paleocene-Eocene to drifts construction-dominated driven by alongslope circulation starting in the Eocene	Eocene-Oligocene boundary. 33–32 Myr
AR3					Cretaceous-Tertiary boundary. Base of Cenozoic sequences
					Upper Aptian
				Upper boundary of continental volcanic-magmatic episodes related to break-up (Seward Deeping Reflectors)	Hauterivian ($\sim$125 Myr)

References

Aceñolaza FG (2000) La Formación Paraná (Mioceno medio): estratigrafía, distribución regional y unidades equivalentes. In: Aceñolaza FG, Herbst R (eds) El Neógeno de Argentina. INSUGEO, Tucumán, Serie Correlación Geológica 14:9–27

Argentina Hydrographic Survey (1976) Bathymetric Chart Western South Atlantic. Argentina Hydrographic Survey, Buenos Aires

Barker PF (1972) A spreading centre in the east Scotia Sea. Earth Planet Sci Lett 15(4):123–132

Barker PF, Dalziel IWD (1983) Progress in geodynamics in the Scotia Arc region. In: Geodynamics of the eastern Pacific region, Caribbean and Scotia Arcs. American Geophysical Union, Geodynamics Series 9:137–170

Barker PF, Dalziel IWD, Storey BC (1991) Tectonic development of the Scotia Arc region. In: Tingley RJ (ed) The geology of Antarctica. Clarendon Press, Oxford, 6:215-248

Bassetto M, Alkmim FF, Szatmari P, Mohriak WU (2000) The oceanic segment of the southern Brazilian margin: morpho-structural domains and their tectonic significance. In: Mohriak W, Talwani M (eds) Atlantic rifts and continental margins. American Geophysical Union, Washington D.C., Geophysical Monograph 115:235–259

Bender VB, Hanebuth TJJ, Chiessi CM (2013) Holocene shifts of the Subtropical Shelf Front off southeastern South America controlled by high and low latitude atmospheric forcings. Paleoceanography 28:1–10

Berger WH, Wefer G (1996) Central themes of South Atlantic circulation. In: Wefer G, Berger WH, Siedler G, Webb DJ (eds) The South Atlantic, present and past circulation. Springer, pp 1–11

Brett CP (1977) Seismicity in the South Sandwich Islands region. J Roy Astron Soc 51(2):453–464

Bryden HL, Imawaki S (2001) Ocean heat transport. In: Siedler G, Church J, Gould J (ed) Ocean circulation and Climate. Elsevier New York: 455–474

Carter L, Cortese G (2009) Change in the Southern Ocean: responding to Antarctica. In: Brigham-Grette J, Powell R, Newman L, Kiefer T (eds) PAGES News: change at the Poles, a paleoscience perspective, IGBP-PAGES (Past Global Changes), PAGES International Project Office 17(1):30–32

Cavallotto JL, Violante RA, Hernández-Molina FJ (2011) Geological aspects and evolution of the Patagonian continental margin. Biol J Linn Soc 103:346–362

Codignotto JO, Kokot RR, Marcomini SC (1993) Desplazamientos verticales y horizontales de la costa argentina en el Holoceno. Revista de la Asociación Geológica Argentina 48:125–132

Dalziel IDW (1982) The evolution of the Scotia Arc, a review. In: Oliver RL, James PR, Jago JB (eds) Antarctic earth science. Cambridge University Press, pp 283–288

Dalziel IWD, Elliot DH (1971) Evolution of the Scotia arc. Nature 233:246–252

Dalziel IWD, de Wit MJ, Palmer KF (1974) Fossil marginal basin in the southern Andes. Nature, London 250:291–294

Dalziel IWD, Lawver LA, Norton IO, Gahagan LM (2013) The Scotia Arc: genesis, evoklution, global significance. Earth Planet Sci Annu Rev 41:767–793

de Witt MJ (1977) The evolution of the Scotia Arc as a key for the reconstruction of the southwestern Gondwanaland. Tectonophysics 37:53–81

de Witt MJ, Jeffrey M, Bergh H, Nicolaysen L (1988) Geological map of sectors of Gondwana. Scale 1:10.000.000. American Association of Petroleum Geologists and University of Witwatersrand. Tulsa

Díaz HF, Markgraff V (1992) El Niño: historical and paleoclimatic aspects of the Southern Oscillation. Cambridge University Press, Cambridge, 96 pp

Duplessy JC, Shackleton NJ, Fairbanks RG, Labeyrie L, Oppo D, Kallel N (1988) Deepwater source variations during the last climatic cycle and their impact on the global deepwater circulation. Paleoceanography 3:343–360

Einsele G (2000) Sedimentary basins, evolution, facies and sediment budget, 2nd ed. Springer, Berlin, 792 p

Emery KO, Uchupi E (1984) The geology of the Atlantic Ocean. Springer, New York, Inc., 1050 pp

Etheridge MA, Lister GS, Symonds PA (1989) Application of the detachment model to reconstruction of conjugate passive margins. In: Tankard AJ, Balkwill HR (eds) Extensional tectonics and stratigraphy of the North Atlantic margins. American Association of Petroleum Geologists, Memoir 46:23–40

Ewing M, Lonardi A (1971) Sediment transport and distribution in the Argentine Basin. 5: Sedimentary structure of the Argentine margin, basin and related provinces. In: Ahrens LH, Press F, Runkorn SK, Urey HC (eds) Physics and Chemistry of the Earth. Pergamon Press, London 8:125–249

Ewing M, Eittreim SL, Ewing JI, Le Pichon X (1971) Sediment transport and distribution in the Argentine Basin. 3: Nepheloid layer and processes of sedimentation. In: Ahrens LH, Press F, Runkorn SK, Urey HC (eds) Physics and Chemistry of the Earth. Pergamon Press, London, 8:49–78

Flower BP, Oppo DW, McManus JF, Venz KA, Hodell DA, Cullen JL (2000) North Atlantic intermediate to deep water circulation and chemical stratification during the past 1 Myr. Paleoceanography 15(4):388–403

Franke D, Neben S, Schreckenberger B, Schulze A, Stiller M, Krawczyk CM (2006) Crustal structure across the Colorado Basin, offshore Argentina. Geophys J Int 165:850–864

Franke D, Neben S, Ladage B, Schreckenberge B, Hinz K (2007) Margin segmentation and volcano-tectonic architecture along the volcanic margin off Argentina/Uruguay, South Atlantic. Mar Geol 244:46–67

Franke D, Ladage B, Schnabel M, Schreckenberger B, Reichert C, Hinz K, Paterlini CM, de Abelleyra J, Siciliano M (2010) Birth of a volcanic margin off Argentina, South Atlantic. Geochem Geophys Geosyst 11(2):Q0AB04. doi:10.1029/2009GC002715

Frankel A, McCann W (1979) Moderate and large earthquakes in the South Sandwich Arc. Indicators of tectonic variation along a subduction zone. J Geophys Res 84(B10):5571–5577

Ghidella ME, Paterlini M, Kovacs LC, Rodríguez G (1995) Magnetic anomalies on the Argentina continental shelf. 4th International Congress of the Brazilian Geophysical Society and 1st Latin American Geophysical Conference, Rio de Janeiro. Expanded abstracts: 269–272

Ghidella ME, Lawver LA, Gahagan LM (2007) La Ruptura de Gondwana y la Apertura del Atlántico Sur: Revisión de Modelos de Tectónica de Placas. VI Simposio Argentino y III Latinoamericano sobre Investigaciones Antárticas. Dirección Nacional del Antártico/Instituto Antártico Argentino. Buenos Aires

Gladczenko TP, Hinz K, Eldholm O, Meyer H, Neben S, Skogseid J (1997) South Atlantic volcanic margins. J Geol Soc London 154:465–470

Gruetzner J, Uenzelmann-Neben G, Franke D (2011) Variations in bottom water activity at the southern Argentine margin: indications from a seismic analysis of a continental slope terrace. Geo-Mar Lett 31:405–417

Gruetzner J, Uenzelmann-Neben G, Franke D (2012) Variations in sediment transport at the Argentina continental margin during the Cenozoic. G3 Geochem Geophys Geosyst 13(10):1–15. Q10003, doi:10.1029/2012GC004266

Haq BU, Hardenbol J, Vail PR (1987) Chronology of fluctuating sea levels since the Triassic. Science 235:1156–1167

Hastenrath S (1982) On meridional heat transport in the World Ocean. J Phys Oceanogr 12:922–927

Heezen BC, Menard HW (1966) Topography of the deep-sea floor. In: Hill MN (ed) The sea. Intersciences Publ, 2nd Edn, J Wiley and Sons: 233–280

Hernández-Molina FJ, Paterlini CM, Violante RA, Marshall P, de Isasi M, Somoza L, Rebesco M (2009) A contourite depositional system on the Argentine Slope: an exceptional record of the influence of Antarctic water masses. Geology 37(6):507–510

Hernández-Molina FJ, Paterlini CM, Somoza L, Violante RA, Arecco MA, de Isasi M, Rebesco M, Uenzelmann-Neben G, Neben S, Marshall P (2010) Giant mounded drifts in the Argentina Continental Margin: origins, and global implications for the history of thermohaline circulation. Mar Pet Geol 27:1508–1530

Hinz K (1981) A Hypothesis on terrestrial catastrophes. Wedges of very thick oceanward dipping layers beneath passive continental margins. Geologische Jahrbuch, E 22

Hinz K, Neben S, Schreckenberger B, Roeser HA, Block M, Gonzalvez de Souza K, Meyer H (1999) The Argentina continental margin north of 48°S: sedimentary successions, volcanic activity during breakup. Mar Pet Geol 16:1–25

Isla FI, Rutter NW, Schnack EJ, Zárate MA (2000) La transgresión belgranense en Buenos Aires. Una revisión a cien años de su definición. Asociación Geológica Argentina, Rev. Cuaternario y Ciencias Ambientales, CADINQUA, Publ. Esp. 4(1):3–14

Kay SM, Ramos VA, Dickinson WR (eds) (2009) Backbone of the Americas: shallow subduction, plateau uplift, and ridge and terrane collision. Geological Society of America, Memoir 204, 279 pp

Kennett J (1982) Marine Geology. Prentice Hall, 813 pp

Ladd JW (1974) South Atlantic and sea floor spreading and Caribbean tectonics. Ph.D. Thesis, Columbia University (ined.), New York, 251 pp

Laprida C, García Chapori N, Chiessi CM, Violante RA, Watanabe S, Totah V (2011) Middle Pleistocene sea surface temperature in the Brazil-Malvinas Confluence Zone: Paleoceanographic implications based on planktonic foraminifera. Micropaleontology 57 (2):183–194

Lawver LA, Gahagan LM (2003) Evolution of Cenozoic seaways in the circum-Antarctic region. Palaeogeogr Palaeoclimatol Palaeoecol 198:11–37

Leat PT, Smellie JL, Millar IL, Larter RD (2003) Magmatism in the South Sandwich Arc. In: Larter RD, Leat, PT (eds) Intra-oceanic subduction systems: tectonic and magmatic processes. Geological Society of London, Sp. Publications 219:285–313

Light M, Maslanyi M, Greenwood R, Banks N (1993) Seismic sequence stratigraphy and tectonics offshore Namibia. In: Williams GD, Dobb A (eds) Tectonics and seismic sequence stratigraphy. The Geological Society, London, Special Publication 71:163–191

Lister GS, Etheridge MA, Symonds PA (1991) Detachment models for the formation of passive continental margins. Tectonics 10:1038–1064

Livermore R, Eagles G, Morris P, Maldonado A (2004) Shackleton Fracture Zone: no barrier to early circumpolar ocean circulation. Geology 32:797–800

Lodolo E, Tassone A (2010) Gateways and climate: the Drake Passage opening. Bolletino di Geofisica Teorica ed Applicata 51:77–88

Lonardi A, Ewing M (1971) Sediment transport and distribution in the Argentine Basin. In: Press F, Runkorn SK, Urey HC (eds) Physics and Chemistry of the Earth. Pergamon Press, London 8:253–264

Macdonald D, Gomez-Perez I, Franzese J, Spalletti L, Lawver L, Gahagan L, Dalziel I, Thomas C, Trewin N, Hole M, Paton D (2003) Mesozoic break-up of SW Gondwana: implications for regional hydrocarbon potential of the southern South Atlantic. Mar Pet Geol 20(3–4):287–308

Malumián N (1999) La sedimentación y el volcanismo terciarios en la Patagonia extraandina. La sedimentación en la Patagonia extraandina. In: Caminos R (ed) Geología Argentina. Anales del Instituto de Geología y Recursos Minerales, Buenos Aires 18:557–612

Malumián N, Nañez C (1996) Microfósiles y nanofósiles calcáreos de la plataforma continental. In: Ramos VA, Turic M (eds) Geología y Recursos Naturales de la Plataforma Continental Argentina. XIII Congreso Geológico Argentino y III Congreso de Exploración de Hidrocarburos, Buenos Aires. Asociación Geológica Argentina-Instituto Argentino del Petróleo, Relatorio 5:73–93

Malumián N, Nañez C (2009) The Atlantic upper Cretaceous-Cenozoic transgressions in Patagonia and the Fuegian Andes: foraminifera, paleoecology and paleogeography. Workshop "Paleogeography and paleoclimatology of Patagonia: Effects on Biodiversity", La Plata, Abstracts: 31

Markgraf V (2001) Interhemispheric climate linkages. Academic Press, 454 pp

Mata JL, Campos M, Basso E, Compagnucci RH, Fearnside P, Magrin G, Marengo J, Moreno ASR, Suárez A, Solman S, Villamizar A, Villers L, Argenal F, Artigas C, Cabildo M, Codignotto JO, Confalonieri U, Magaña V, Morales-Arnao B, Oropeza O, Pabón J, Paz J, Paz O, Picado F, Poveda G, Tarazona J, Vargas W (2001) Latin America. In: McCarthy JJ, Canziani OF, Leary NA, Dokken DJ, White KS (eds) Climate change 2001. Impacts, adaptation and vulnerability. Contribution of Working Group II to the 3rd assessment report of the Intergovernmental Panel on Climate Change. Cambridge University Press, pp 693–734

Matano RP, Palma ED, Piola AR (2010) The influence of the Brazil and Malvinas Currents on the southwestern Atlantic shelf circulation. Ocean Sci 6(4):983–995

Max MD, Ghidella M, Kovacs L, Paterlini CM, Valladares JA (1999) Geology of the Argentine continental shelf and margin from aeromagnetic survey. Mar Pet Geol 16:41–64

McAdoo DC, Laxon SW (1997) Antarctic tectonics: constraints from an ERS-1 satellite marine gravity field. Science 276:556–560

Meisling KE, Cobbold PR, Mount VS (2001) Segmentation of an obliquely rifted margin, Campos and Santos basins, southeastern Brazil. AAPG Bull 85(11):1903–1924

Mohriak WU, Rosendahl BR, Turner JP, Valente SC (2002) Crustal architecture of South Atlantic volcanic margins. In: Menzies MA, Klemperer SL, Ebinger CJ, Baker J (eds) Volcanic rifted margins. Geological Society of America, Boulder, Colorado, Special Paper 362:159–202

Mutter JC, Talwani M, Stoffa PL (1982) Origin of seaward-dipping reflectors in oceanic crust off the Norwegian margin by "subaerial sea-floor spreading". Geology 10:353–357

Mutter JC, Buck WR, Zehnder CM (1988) Convective partial melting—a model for the formation of thick basaltic sequences during the initiation of spreading. J Geophys Res 93(B2):1031–1048

Neben S, Schreckenberger B, Adam J, Behrens T, Bönnemann C, Demir Ü, Franke D, Heyde I, Kallaus G, Ladage S, Müller C, Schrader U, Sievers J, Surburg E, Temmler T (2005) Research Cuise BGR04—ARGURU—Geophysical Investigations offshore Argentine & Uruguay with Akademik Alexandr Karpinsky. Cruise Report and Preliminary results. Bundesanstalt für Geowissenschaften und Rohstoffe, Hannover, 99 pp

Oberhansli H, Hsü KJ (1986) Paleocene-Eocene paleoceanography. In: Hsü KJ (ed) Mesozoic and Cenozoic oceans. Geodynamics Ser. 15. Geol Soc Am 85–100, Boulder, Colorado

Parker G, Violante RA, Paterlini CM (1996) Fisiografía de la Plataforma Continental. In: Ramos V, Turic M (eds) Geología y Recursos Naturales de la Plataforma Continental Argentina. Relatorio XIII Congreso Geológico Argentino. Asociación Geológica Argentina, Buenos Aires, 1–16

Parker G, Paterlini CM, Violante RA, Costa IP, Marcolini S, Cavallotto JL (1999) Descripción Geológica de la Terraza Rioplatense (Plataforma Interior del Noreste Bonaerense). Servicio Geológico y Minero Argentino, Boletín N° 273: 98 pp

Parker G, Violante RA, Paterlini CM, Marcolini S, Costa IP, Cavallotto JL (2008) Las secuencias sismoestratigráficas del Plioceno-Cuaternario en la Plataforma Submarina adyacente al litoral del este bonaerense. Lat Am J Sedimentol Basin Anal 15:105–124

Pelayo AM, Wiens DA (1989) Seismotectonics and relative plate motion in the Scotia Sea region. J Geophys Res 94(B6):7293–7320, Washington

Potter PE, Szatmari P (2009) Global Miocene tectonics and the modern world. Earth Sci Rev 96 (4):279–295

Preu B, Schwenk T, Hernández-Molina FJ, Violante RA, Paterlini CM, Krastel S, Tomasini J, Spieß V (2012) Sedimentary growth pattern on the northern Argentine slope: the impact of North Atlantic Deep Water on southern hemisphere slope architecture. Mar Geol 329–331:113–125

Rabassa J, Coronato AM, Salemme M (2005) Chronology of the Late Cenozoic Patagonian glaciation and their correlation with biostratigraphic units of the Pampean region (Argentina). J S Am Earth Sci 20:81–103

Rabinowitz PD, LaBreque JL (1979) The Mesozoic South Atlantic and evolution of its continental margins. J Geophys Res 84(B11):5973–6002

Rahmostort S (2006) Thermohaline ocean circulation. In: Elias SA (ed) Encyclopedia of Quaternary sciences. Elsevier, Amsterdam, 10 pp

Ramos VA (1996) Evolución tectónica de la plataforma continental. In: Ramos VA, Turic MA (eds) Geología y Recursos Naturales de la Plataforma Continental Argentina. XIII Congreso Geológico Argentino y III Congreso de Exploración de Hidrocarburos, Buenos Aires. Asociación Geológica Argentina-Instituto Argentino del Petróleo, Relatorio 21:405–422

Ramos VA (1999) Rasgos Estructurales del Territorio Argentino. Evolución Tectónica de la Argentina. In: Caminos R (ed) Geología Argentina. Instituto de Geología y Recursos Minerales, SEGEMAR. Anales 29(24):715–784

Ramos VA, Turic M (eds) (1996) Geología y Recursos Naturales de la Plataforma Continental Argentina. XIII Congreso Geológico Argentino, Buenos Aires

Ramos VA, Riccardi AC, Rolleri EO (2004) Límites naturales del norte de la Patagonia. Revista Asociación Geológica Argentina 59:785–786

Rapalini AE (2007) Paleomagnetic test of the Patagonian Orocline. Geol Acta 5(4):287–294

Rostami K, Peltier WR, Manzini A (2000) Quaternary marine terraces, sea-level changes and uplift history of Patagonia, Argentina: comparisons with predictions of the ICE-4G (VM2) model of the global process of glacial isostatic adjustment. Quatern Sci Rev 19:1495–1525

Sandwell DT, Smith WHF (1997) Marine gravity anomaly from Geosat and ERS 1 Satellite altimetry. J Geophys Res 102:10039–10054

Sarnthein M, Winn K, Jung JJA, Duplessy JC, Labeyrie L, Erlenkeuser H, Ganssen G (1994) Changes in east Atlantic deepwater circulation over the last 30,000 years: eight time slice reconstructions. Paleoceanography 9:209–267

Schümann TK (2002) The hydrocarbon potential of the deep offshore along the Argentine volcanic rifted margin–A numerical simulation, PhD thesis, RWTH Aachen Univ, Aachen, Germany, 194 pp

Soto M, Morales E, Veroslavsky G, de Santa Ana H, Ucha N, Rodríguez P (2011) The continental margin of Uruguay: crustal architecture and segmentation. Mar Pet Geol 28:1676–1689

Talley LD (2003) Shallow, intermediate and deep overturning components of the global heat budget. J Phys Ocean 33:530–560

Tankard AJ, Uliana MA, Welsink HJ, Ramos VA, Turic M, Franca AB, Milani EJ, Brito Neves BB, Eyles N, Skarmeta J, Santa Ana H, Wiens F, Cibrián M, López Paulsen O, Germs GJB, de Wit MJ, Machacha T, Miller RMcG (1995) Structural and tectonic controls of basin evolution in southwestern Gondwana during the Phanerozoic. In: Tankard AJ, Suárez Soruco R, Welsink HJ (eds) Petroleum basins of South America. American Association of Petroleum Geologists Memoir 62:5–52, Tulsa

Thomson K (1998) When did the Falklands rotate? Mar Pet Geol 15:723–736

Turic MA, Nevistic AV, Rebay G (1996) Geología y Recursos Naturales de la Plataforma Continental. In: Ramos V, Turic M (eds) Geología y Recursos Naturales de la Plataforma Continental Argentina. Relatorio 13° Congreso Geológico Argentino. Buenos Aires, pp 405–423

Urien CM, Zambrano JJ (1996) Estructura de la Plataforma Continental. In: Ramos VA, Turic M (eds) Geología y Recursos Naturales de la Plataforma Continental Argentina. XIII Congreso Geológico Argentino y III Congreso de Exploración de Hidrocarburos, Buenos Aires 1996. Asociación Geológica Argentina-Instituto Argentino del Petróleo, Relatorio 1:29–66

van Andel TjH, Thiede J, Sclater JG, Hay WW (1977) Depositional history of the South Atlantic during the last 125 million years. J Geol 85:651–698

Violante RA (2003) Geomorphology and coastal evolution during the last interglacial maximum sea-level highstand in the Atlantic littoral of the Pampean region, Argentina. "Puglia 2003" Final Conference Project IGCP 437 "Coastal Environmental Change during Sea-Level Highstands", Otranto-Taranto, Italia. Abstracts: 243–245

Violante RA, Parker G (1993) Evidencias de una ingresión marina intercalada entre el Interensenadense y el Belgranense ("Pampeano") en las áreas costeras y marinas del este bonaerense. XII Congre-so Geológico Argentino, Mendoza. Actas II:252–259

Violante RA, Parker G (1999) Historia evolutiva del Río de la Plata durante el Cenozoico superior. XIV Congreso Geológico Argentino, Salta. Actas 1:504–507
Violante RA, Paterlini CM, Bozzano G, Hanebuth TJJ, Strasser M, Krastel S (2010a) Morphosedimentary characteristics and evolution of the Mar del Plata Submarine Canyon, Northern Argentina Continental Margin. 18° International Sedimentological Congress, Mendoza, Argentina, Abstracts: 907
Violante RA, Paterlini CM, Costa IP, Hernández-Molina J, Segovia LM, Cavallotto JL, Marcolini S, Bozzano G, Laprida C, García Chapori N, Bickert T, Spiess V (2010b) Sismoestratigrafía y evolución geomorfológica del talud continental adyacente al litoral del este bonaerense, Argentina. Lat Am J Sedimentol Basin Anal 17:33–62
Voigt I, Chiessi CM, Prange M, Mulitza S, Groeneveld J, Varma V, Henrich R (2015) Holocene shifts of the southern westerlies across the South Atlantic. Paleoceanography 30(2):39–51
Wefer G, Berger WH, SiedlerG, Webb DJ (eds) (1996) The South Atlantic, present and past circulation. Springer, 644 pp
Wefer G, Mulitza S, Ratmeyer V (eds) (2004) The South Atlantic in the Late Quaternary. Springer, 722 pp
Yrigoyen MR (1979) Cordillera principal. II Simposio de Geología Regional Argentina 1, Actas: 651–649
Yrigoyen MR (1999) Los depósitos cretácicos y terciarios de las Cuencas del Salado y Colorado. In: Caminos R (ed) Geología Argentina. Servicio Geológico Minero Argentino, Instituto de Geología y Recursos Minerales, Anales 29:645–650
Zachos J, Pagani M, Sloan L, Thomas E, Billups K (2001) Trends, rhythms and aberrations in global climate 65 Myr to present. Science 292:686–693
Zambrano JJ, Urien CM (1974) Pre-Cretaceous basins in the Argentine continental shelf. In: Burk CA, Drake CL (eds) The geology of continental margins. Springer, pp 463–470

Chapter 4
Morphosedimentary Configuration of the Argentina Continental Margin

Abstract The major regional features of the margin are coastal plains, shelf, slope and rise. These features show varied characteristics depending on their location on each type of margin. They contain diverse subordinated morphosedimentary features related to the prevailing genetic processes acting on each of them and shaping their relief. The shelf is shaped in terraces at increasing stepping depths toward offshore, they having been genetically related to fluctuations in the sea-level rise during the post-glacial transgression. The slope is also shaped in terraces, but in this case, these features are genetically associated to the development of alongslope complex systems of mixed depositional and erosive contouritic features formed at different depths, possibly related to highly energetic interfaces between adjacent water masses that constitute the thermohaline oceanic system. Gravitational, downslope processes generating turbidites, mass transport and debris flows deposits, most of them acting inside or close to submarine canyons, actively interact with the contouritic processes. The rise is partially formed by gravity-driven deposits at the base of the slope, although in the southern part of the margin the alongslope, contouritic processes are strong enough to shape the rise and imprinting it with particular current-driven features.

Keywords Argentina Continental Margin · Morphosedimentary configuration · Coastal plains · Shelf · Slope · Rise · Submarine canyons · Argentine basin · Contourites · Turbidites

The ACM covers an area of $\sim 2 \times 10^6$ km^2 along different geotectonic settings (as described above, Fig. 3.4). The morphosedimentary features constituting the margin are the result of a complex set of variables such as the characteristics inherited from its evolution, sedimentary processes and dynamics, and oceanographic conditioning factors. The margin is composed of typical features of fourth order according to Heezen and Menard (1966), such as a wide shelf, slope and rise, as

R.A. Violante et al., *The Argentina Continental Margin*,
SpringerBriefs in Earth System Sciences, DOI 10.1007/978-3-319-04196-4_4

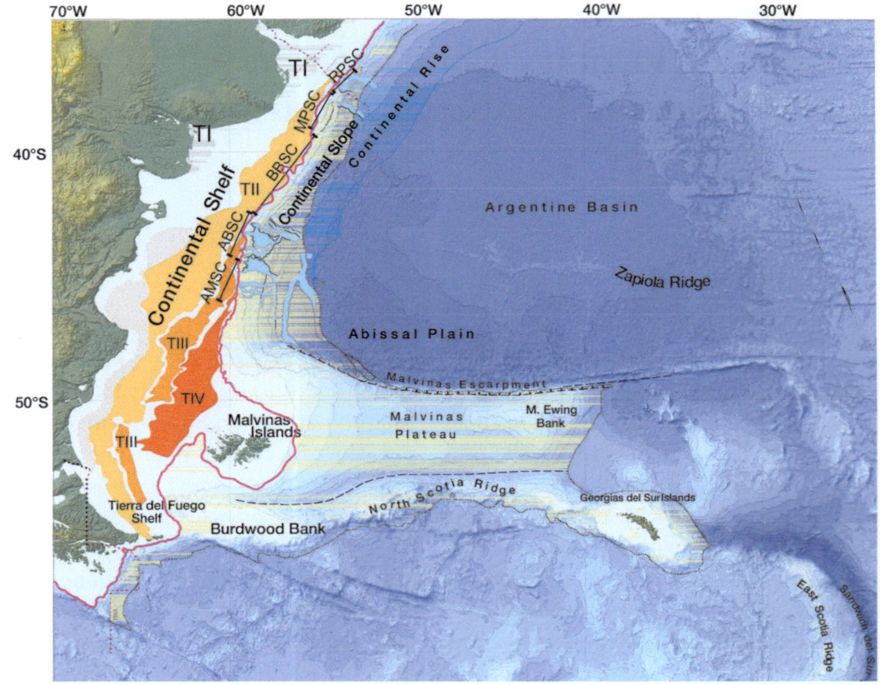

Fig. 4.1 Major morphosedimentary features in the Argentina Continental Margin (modified after Cavallotto et al. 2011, who compiled the maps from Parker et al. 1996, 1997 for the continental shelf, and Hernández-Molina et al. 2009 for the slope). The red line represents the shelf outer edge. TI, II, III and IV: terraces in the shelf (see text). RPSC: Rio de la Plata submarine canyon system. MPSC: Mar del Plata submarine canyon system. BBSC: Bahía Blanca submarine canyon system. ABSC: Almirante Brown submarine canyon system. AMSC: Ameghino submarine canyon system

well as submarine canyons (Figs. 3.3 and 4.1). Following Kennett (1982) and in the sense that it is considered in this contribution, coastal plains are also included as a part of the continental margin.

4.1 Coastal Plains

Although not always described as a component of the continental margins, the coastal plains are a feature of a major importance as they represent the "present" picture of a complex regional evolution occurred through very long periods of time dominated by back and forth shifting of the shoreline as a result of interaction among different marine and continental processes (relative sea-level fluctuations, isostasy, tectonism, subsidence/uplift, inherited relief from endogene processes, and also exogene processes like fluvial, aeolian, littoral—waves, tides and coastal currents—etc.). As a result of that shifting along its history, successive shorelines were set (and preserved both inland and offshore) at each position where the combination of processes momentaneously enabled sea-level to stood and remain for a certain period of time. The relicts of ancient shorelines formed during lower-than-today sea-level positions are part of the continental shelf and therefore they are presently drowned by the sea. On the other hand, ancient shorelines were formed during higher-than-today sea-level positions, so constituting presently subaerially exposed "marine" features, which considered as a whole represent the region affected by the retreat in sea level, i.e., the "coastal plains." Therefore, coastal plains undoubtedly represent the marginal environments of continental margins. The concept of coastal plains is normally restricted to the set of coastal relict environments evolved during the last (Late Holocene) sea-level retreat.

The Argentina's coast, which exceeds 5000 km long, has a regional NE–SW orientation inherited from the seafloor expansion and separation between the South America and Africa Plates. However, its shape and geological composition are also derived from the geological constitution of the substratum, the existence of transverse-to-the-shore basins and headlands, and the geomorphic features characterizing the continent. As a result, two major types of coasts are present. In the north, adjacent to the Pampean region (from Rio de la Plata at 34°S to Negro river at 41°S) (for coastal geographic locations see Fig. 3.3), the Bonaerensian coasts develop with alternation of lowlands (large deltaic and estuarine coasts, tidal flats, and beach/dunes associated to ancient coastal barriers–lagoon systems) and cliffs (low height, made up of soft and semiconsolidated loessic sediments). On the other hand, adjacent to the Patagonian region (south of 41°S) high-cliffed coasts (up to 100 m high) composed of more consolidated sandy and gravelly sediments as well as rocky outcrops—most of them developing wave-cut platforms at the present sea level—are the most significant features, except around the outlet of the major rivers where some small estuarine and coastal barriers environments develop.

The low coastal areas are associated to coastal plains, which contain the records of ancient shorelines and related environments developed during the Late Holocene, when sea level was—for the last time—higher than today before it fell down to its present position. In this way, former beach ridge plains, coastal barriers/lagoons, estuaries and deltas can be recognized in areas presently located far inland from the shoreline, as in most of the coastal lowlands of the Pampean region and in restricted sectors of Patagonia. In Patagonia, shorelines formed during highstands in sea level at Plio-Pleistocene times have been later subjected to isostatic uplift and rised, presently constituting "terraces" at progressively higher altitude to the south. They were described by a large number of authors (e.g., Feruglio 1950; Codignotto 1987; Codignotto et al. 1992, 1993; Rostami et al. 2000; Schellmann and Radtke 2000; Isla and Bujalesky 2008; Pedoja et al. 2011, among others), who profusely discussed the origin and tectonic–isostatic implications in their present position.

It is worth mentioning here that the argentine coastal plains, as part of the Southern Hemisphere, show, unlike the coastal plains from the Northern Hemisphere, evidences of higher-than-today Late Holocene sea levels (Isla 1989).

4.2 Continental Shelf

It covers an area of $\sim 1 \times 10^6$ km^2 (Figs. 3.3 and 4.1). Its width increases to the south in the passive margin sector, with ~ 170 km in the northern region to ~ 850 km in the south at the latitude of Malvinas Islands. In the active margins of the south, the shelf is extremely narrow reaching only few kilometers south of de los Estados Island. According to Lonardi and Ewing (1971), the Burdwood Bank might be considered a part of the shelf. The surface of the entire shelf is smooth with little and low irregularities not higher than 20 m.

A major distinctive feature of the shelf (Parker et al. 1996, 1997; Violante et al. 2014b) is the configuration of the 100 m isobath (Fig. 3.3), which runs very close (only few tens of km apart) to the shelf break north of 43°S and hence separated up to 500 km from the coast. However, south of 43°S that isobath progressively approaches the coast reaching distances from it not larger than 100–200 km at 47° S, and therefore is separated 500 km from the shelf break. In relation to this, the shape of the shelf strongly changes in cross section, as in the northern areas it is convex (steeper surface in the outer shelf than in the inner shelf), whereas in the southern areas it is concave (steeper surface in the inner shelf than in the outer shelf). This configuration is considered, following the concepts by Lonardi and Ewing (1971), to be the result of the swept of the shelf surface by the northward-flowing Malvinas current which is stronger in the south, at the time that it deflects to the east as approaches the Brazil–Malvinas Confluence.

Apart from this regional configuration, the shelf surface in the passive sector of the ACM is modeled in four major terraces with top subhorizontal surfaces at 25/30 m (T1), 85/95 m (T2), 110/120 m (T3) and 130/150 m (T4), separated to each other by steps represented by strongly steep surfaces (Fig. 4.1). The terraces (that were mentioned for the first time in the mid 20th century by Groeber 1948) are oriented in a nearly NNE–SSW direction parallel to the general trend of the shoreline and are not uniformly distributed. Sediments on the terraces' surface are mostly relictic Holocene sands with subordinate gravels and muds. Parker et al. (1996), Violante (2005), Perillo and Kostadinoff (2005), Ponce et al. (2011) and Violante et al. (2014b, e) provide details of the terraces characteristics. The terraces extend northwards along the Uruguayan and Brazilian shelves. In Brazil they have been described by several authors (Correa 1996; Conti 2004; Mahiques et al. 2010; among others), in most of the cases its origin having been associated to post-glacial sea-level fluctuations.

T1 is the northernmost terrace; it extends offshore the Pampean region along 900 km (from 35° to 42°S), although is interrupted and splitted in two parts in its middle section at the latitude of the Mar del Plata city (Corrientes cape, 38°S). There, a low mountain chain (Tandilia ridge) composed of Precambrian and Paleozoic rocks extends from the continent, reaches the coast as low-relief hills, and continues under the shelf surface partially interrupting the submarine features (including the terrace). The terrace is 150 km wide at the mouth of the Río de la Plata and 100 km in front of the bay of Bahia Blanca. Bathymetric information indicates that it probably continues in a NNE direction along the coasts of Uruguay. The northern part of T1 at the mouth of Rio de la Plata, named "Rioplatense Terrace" (Parker et al. 1999; Violante and Parker 2000, 2004), depicts a subhorizontal top surface approximately triangular in shape, with a maximum width of 150 km and a maximum length of 500 km in its outer border at 30 m water depth. Slope gradient of the surface is gentle (1:10,000), slightly dipping southeastward, and is bounded by two steps of steeper slope gradient (1:500 to 1:1000), which correspond to the shoreface in its landward side at 10 m water depth, and the outer step that separates the terrace from the outer shelf, whose base extends between Mar del Plata (Argentina) and Punta del Este (Uruguay) reaching 70/80 m water depth. The outer step has a non-uniform surface with three smaller, second-order terraces at water depths of 50, 70 and 80 m. The surface of the terrace is mantled by a relictic and palimpsestic Holocene sandy blanket made up of fine–to-medium sands containing shell fragments and minor amounts of coarse sands and pebbles of sandstones, quartzites and caliche. The coarsest sediments, frequently bioclastic, are concentrated along lineal patches showing a NE–SW orientation subparallel to the present shoreline and isobaths, indicating the position of ancient beaches (Lonardi and Ewing 1971; Urien and Ewing 1974; Violante and Parker 2000, 2004; Parker et al. 1999, 2008; Violante et al. 2014b). One of these features is the so-called "La Plata Shoal," at the outer edge of the terrace at 30 m water depth, which shows a conspicuous ridge-like morphology modeled over a substratum (nucleous of the bank) of probable Late Pleistocene (Sangamonian) age (Parker et al. 2008; Isla and Madirolas 2009). The sediment blanket that covers the terrace resulted from the landward migration of littoral barrier systems evolved during the post-LGM transgression and the consequent reworking of the sediments that remained submerged as a result of the ravinement process that accompanied the shoreline retreat, as well as reworking of the pre-Holocene substratum (Urien and Ewing 1974; Parker et al. 1999; Violante and Parker 2000, 2004). On the other hand, the southern part of T1, offshore Bahia Blanca, is a feature covered by medium and coarse sands containing pebbles in front of the outlet of the Colorado and Negro rivers. Spalletti and Isla (2003) and Melo et al. (2003, 2013), and later Violante et al. (2014a) in a more regional context, described in the region the presence of submerged deltaic facies developed during sea-level lowstands, which are here interpreted as the primary modelers of this sector of the terrace before the final reworking by marine processes.

TII is the largest terrace of the shelf as it extends along 2500 km between Río de la Plata and Tierra del Fuego (from 35° to 55°S), with a maximum width of 200 km. Bathymetric information reveals that this terrace shows the presence of second-order steps below 90 m water depth. The depth of 90 m is not only a morphological boundary along the entire shelf but also a sedimentological boundary, since above that depth silty clayey fine sands dominate, whereas below that depth (affecting also terraces TIII and TIV) medium and coarse sands with some gravel concentrations are present. It would be probably related to the fact that during times of low sea level, sediment dynamic was greater as a result of larger fluvial activity because of the increased surface gradient resulting from the lowering in rivers base-level. The outer edge of TII is in some places characterized by relicts of lowstand deposits composed of sedimentary sequences representing former coastal environments of fluvial–estuarine, barriers and coastal lagoons developed immediately after the LGM (Violante et al. 2014d).

TIII extends offshore the southern Patagonia (from 43° to 54°S), with a width not exceeding 100 km. Its surface is characterized by relictic features that represent a partially buried fluvial network where the paleodrainage pattern can be recognized (Parker et al. 1996; Ponce et al. 2011; Violante et al. 2014e). Coarser sediments, mainly gravels, are considered as relicts of glacifluvial deposits carried to the sea during glacial times.

TIV is the smallest terrace having 700 km long and up to 150 km wide, located west of Malvinas Islands (from 46° to 52°S). As in TIII, gravelly deposits are considered as of glaciofluvial origin and a relictic fluvial network is also observed on its surface.

The shelf substratum is composed of different Neogene sedimentary sequences that partially crop out at the steps between adjacent terraces, being the outcropping units younger as we approach to the shallowest terraces. Those sequences are of continental and marine origin, and the uppermost ones constitute the records of the Quaternary glacieustatic marine transgressions (Violante and Parker 2000, 2004; Parker et al. 2008).

The ancient fluvial network partially preserved in the shelf surface is considered to have developed during the last lowstand when the shelf was exposed to subaerial conditions. This paleo-fluvial network has not been completely obliterated during the post-glacial transgression, so as the paleodrainage pattern can be recognized (Parker et al. 1996; Violante et al. 2007, 2014e). Although in a regional sense the drainage pattern is dendritic, north of 42°S the valleys are subparallel and regionally oriented to the southeast. South of 42°S, the pattern is more chaotic with numerous distributaries; main valleys are oriented to the south and then change to the southeast as approaching their lowermost parts. Between 42 and 46°S some valleys seem to be connected with submarine canyons. South of 46°S the valleys tend to converge toward the Malvinas Trough. In relation to relict fluvial networks in coastal semi-enclosed areas, they show particular patterns like a more irregular branching of tributary valleys associated to submerged deltas (like the Colorado–Negro deltaic system), or a semiradial, centripetal drainage in semi-enclosed basins (San Matías, Nuevo and San Jorge gulfs). Patches of coarse sands with a significant gravel content offshore the outlet of the Colorado river are considered as relicts of presently submerged deltaic facies (Spalletti and Isla 2003; Melo et al. 2003, 2013; Violante et al. 2014a). Ponce et al. (2011) stated that the larger extension of the land (emerged regions) during glacial times favored a better distribution and integration of the drainage network. Other indicators of subaerial exposure of the shelf are relicts of continental sediments including paleosoils that have been recorded in cores obtained in the northernmost regions of the shelf (Osterrieth et al. 2008), as well as findings of mammal remains in nearby areas (Cione et al. 2005) and vegetation remains in San Matias Gulf (Isla 2011).

Sediments: The sedimentary mantle blanqueting the shelf surface is made up of bioclastic fine sands with subordinated amounts of medium sands, shell fragments, gravels and muds. In general, these sediments are relictic as they resulted from reworked coastal and marine deposits formed during the rise and fall of sea level during transgressive–regressive events, partially preserved as morphosedimentary features representing former coasts evolved during the different eustatic cycles. The processes that intervened in the sediment transfer from the continent to the sea have been interpreted for the Patagonian region as being 56% from coastal erosion, 41% from eolian supply, and 3% from fluvial supply (Gaiero et al. 2003; Violante et al. 2014e). In the northern Pampean region it would be probably different if the enormous amount of sediments provided by the Río de la Plata is considered (57–130 Mt/yr, Depetris and Griffin 1968; Depetris and Paolini 1991; Gaiero et al. 2004; Campos et al. 2008). Anyway, and considering that the processes originated in Patagonia are those that affect more the entire shelf as a consequence of the northward-prevailing littoral drift in the nearshore regions as well as the northward-prevailing flow of deeper marine currents—added to the fact that the influence of the Rio de la Plata only affects the regions located north of 38°S—undoubtedly the coastal erosion and the erosive consequences of the swept of the shelf during the Quaternary sea-level fluctuations is a major process involved in the sediment supply and distribution on the shelf. As a consequence, the dominant sediment fraction preserved on the shelf surface is sand, covering approximately 65% of the shelf surface, being mostly fine- and medium-sized with subordinate coarse and very fine fractions. They are clean, moderately well sorted and yellowish, brown and gray-colored, that became darker when containing subordinate mud. The finest fractions (fine and very fine sand) dominate in the central and north Patagonian part of the shelf. Sandy coarse fractions are also abundant in the south Patagonia–Tierra del Fuego sectors. A possible explanation for the lack of fine sediments (fine sands, silts and clays) particularly in the southern Patagonian regions of the shelf is related to the strong northward-flowing currents acting there (possibly enhanced during times of less water covering of the shelf close to glacial times) that prevents their deposition and helps in the delivery of those sediments toward the adjacent Argentine basin where they are deposited in large mudwave fields (Flood et al. 1993; Manley and Flood 1993; von Lom-Keil et al. 2002). In coastal semi-enclosed regions (Río de la Plata, Bahía Blanca, San Matías and San Jorge Gulfs), very fine sands in transition to silts are frequent. In some parts of the shelf, in particular in nearshore areas, sands are modeled in subaqueous dunes, lineal shoals, shoal retreat massifs, barriers and ridges as a result of the dominance of activity of waves or tidal currents and the heritage of ancient morphologies associated to ocean dynamics and sea-level fluctuations.

Bioclasts constituted by shell fragments of pelecipods, brachiopods, foraminifera and ostracods, as well as arthropods, fishes and echinoderms fragments, are essential components of the coarser sand fraction, particularly in the northern regions, where they account for 12.5% of the total sediment composition.

Gravels (in an amount of 12.5% over the entire shelf surface) are mostly abundant in the southern Patagonia shelf and at the mouth of the main Patagonian rivers, associated to glaciofluvial deposits related to the glaciers activity that dominated in large parts of southern Patagonia and Tierra del Fuego during glaciations, including the evidences of presently submerged glacial deposits that have been described offshore Tierra del Fuego linked to the outlet of Magellan Strait (Isla and Schnack 1995; Coronato et al. 1999; Mouzo 2005).

Muds (silts and clays), representing 8% of the shelf surface, are dominant in semi-enclosed coastal areas of restricted circulation, as bays, estuaries and river mouths, as well as in few places of the outer shelf where relicts of low-energy paleoenvironments are preserved. Muds are normally dark green or gray, partially cohesive, and contain high amounts of organic matter.

Outcrops of pre-Cenozoic rocks and Cenozoic consolidated sediments appear in scattered and isolated places where the Quaternary sedimentary cover is absent. These outcrops account for only 2% of the shelf surface.

Geomorphologic provinces and regions of the shelf: the interaction among the different global, regional and local factors (sea-level fluctuations, climate, terrigenous sediment supply, littoral and ocean dynamic, coastal and submarine geomorphic features, erosion-deposition rate, isostasy, tectonism) influenced the shelf evolution and configuration. Depending on the different areas affected and the relative influence of each factor, some regions can be distinguished.

From the sediment dynamic viewpoint three regions are recognized: inner, middle and outer shelf. The ***inner shelf*** develops between the shoreface (its base located at ~10 m water depth) and 30 m water depth. Sediments present there are adjusted to the present-day hydrodynamic conditions influenced by littoral sedimentary processes, coastal currents, as well as waves and tides action; therefore the resulting sediments are considered "palimpsests." Active features are present as shoal retreat massifs, lineal shoals and giant subaqueous dunes. This environment is better defined in the northernmost part of the shelf (Bonaerensian shelf). The ***middle shelf*** extends between 30 and ~90 m water depth; it is represented by areas relatively stable in terms of sediment mobility and hence the sedimentary cover can be considered as relictic. This environment is better defined in the Patagonian shelf. The ***outer shelf*** extends at depths deeper than 90 m reaching the shelf break; sediment dynamics is here associated to the sedimentary processes at the shelf–slope transition, mainly dominated by deep alongslope currents as well as across-slope processes connecting the shelf edge with the head of submarine canyons.

On the other hand, Parker et al. (1997) defined six "Geomorphologic Provinces." The ***Rioplatense Terrace*** constitutes the northern sector of Terrace I adjacent to the Rio de la Plata, shaped during the Holocene in diverse erosive and depositional features and draped by the transgressive sands, formed by relictic sediments later adjusted to the present hydrodynamic conditions and transformed in palimpsests (Parker et al. 1999; Violante and Parker 2000, 2004; Violante 2005). The ***Deltaic front of the Colorado and Negro rivers*** belongs to the southern sector of Terrace I; it is associated to the deltas formed during different stages of the Late Pleistocene–

Holocene sea-level fluctuations (Spalletti and Isla 2003; Melo et al. 2003, 2013; Violante et al. 2014a), evidenced in diverse lobate-shaped features composed of coarse sands, gravels and bioclasts. The ***North Patagonian gulfs*** constitute coastal semi-enclosed basins that reach depths deeper than the outer shelf border, connected to the open sea through shallow sills; its origin was attributed to former continental depressions shaped by eolian and fluvial processes which were later invaded by the sea during the sea-level transgressions (Mouzo et al. 1978; Paterlini and Mouzo 2013; Mouzo 2014). The ***Inner Patagonian Shelf*** extends between Nuevo Gulf and Santa Cruz river at depths equivalent to the base of Terrace I around 70–80 m water depth; despite its relatively smooth surface, some features manifested by marked reliefs and lobate-shaped morphologies are probably associated to little deltaic environments at the mouth a many Patagonian rivers. The ***Outer Patagonian Shelf*** is the largest province, comprising terraces II, III and IV. In most of its surface relicts of an ancient drainage system are evidenced, revealing the subaerial fluvial processes that occurred during pre-transgressive stages. The ***Tierra del Fuego Shelf*** is a surface mostly modeled by glacial and glacifluvial processes during the LGM and associated lowstand; there, submerged moraines and other glacigenic features were recognized (Isla and Schnack 1995; Coronato et al. 1999; Mouzo 2005). The ***Malvinas Islands Shelf*** is a portion of the shelf representing an extension of the Outer Patagonian Shelf.

From the geographic perspective three regions are differentiated. The ***Northern Region***, adjacent to Rio de la Plata, is characterized by the significant influence of a huge fluvial–estuarine environment that was during the Neogene, one of the most important sediment providers to the shelf in that particular region. So, it imprinted this sector with a particular morphosedimentary configuration mostly dominated by fluvial rather than marine processes; relicts of low-energy (estuarine) environments were therefore preserved on the shelf surface with only minor evidences of higher energy environments (coastal plains made up of beaches and sandy barriers). The ***Central Region*** comprises the shelf areas adjacent to the Pampas and central/north Patagonian regions; the region behaved during lowstands as an extensive subaerial plain later flooded by the sea during sea level transgressions; therefore the dominant processes were waves and tides activity with minor fluvial influence, except at the mouth of the major river systems (like Colorado and Negro); glacial influence was inexistent. Extensive coastal plains of barriers and high-energy beaches were reworked as a result of the sweeping of the shelf surface by the ravinement process during the sea-level rise, in such a way that a blanket of relictic and palimpsestic sandy deposits was left on the shelf surface. The ***Southern Region***, in the marine area adjacent to southern Patagonia and Tierra del Fuego, had a completely different behavior during lowstands as the continent is there very narrow and even the present shelf was covered by glacifluvial deposits, including moraines, since glaciers reached positions very close to the present shoreline (Rostami et al. 2000; Coronato et al. 2008; Ponce et al. 2011). As a result, moraine-like morphologies as well as abundant gravels and glacifluvial sediments are common components of this southern region.

4.3 Shelf Break

The shelf break represents the transition from the shelf to the upper slope, where some of the shelf sediments are mobilized downslope mainly by gravitational processes and incorporated to the along—and across-slope processes proper of the continental slope. Pierce and Siegel (1979) estimated a sediment transfer across the shelf break to the slope of 1.7 Mt/yr. The shelf break is variable in depth along the margin, although the general trend is of increasing depths from north to south between 70 and 190 m (Parker et al. 1996; Violante et al. 2014b). Some incisions are evidenced in the shelf break possibly related to ancient river outlets and estuarine environments during stages of sea-level lowstands (Violante et al. 2014d), and only few of them can be associated to upslope extensions of submarine canyon´s heads (Ewing and Lonardi 1971). The places in the southern regions where the shelf break is deeper seem to be related to the location of offshore extensions of tectonic basins and head of submarine canyon systems (Violante et al. 2014b).

4.4 Slope

Traditionally, the continental slope was defined basically on its relief, as the transition zone of steep slope between the shelf border and the ocean floor that is cut by submarine canyons and dominated by gravity downslope processes (Ej.: Kuenen 1950; Shepard 1973; Kennett 1982). However, in the last decades a better understanding of this feature was reached, showing that it is rather characterized by complex morphologies and sets of erosive and depositional features where not only downslope processes take place but also alongslope processes are common.

As one of the largest slope environments in the world, the Argentine slope has different configurations according to the geotectonic setting. In the Volcanic rifted (passive) margin, the slope (herein named PMS—passive margin slope) is a typical continuous, wide, and high-angle feature, extending in a NNE–SSW straight line along 1600 km at water depths ranging from around 120 to 4000 m. On the other hand, in the transcurrent and transpressive margins around the Malvinas Plateau and the North Scotia Ridge, the slope (herein named TMS—transcurrent/transpressive margin slope) shows different features, since along the Malvinas Escarpment and in the North Scotia Ridge (orientations broadly E–W) is very steep and narrow, whereas in the Malvinas Plateau it has varied configurations with extensive and low-angle surfaces.

The PMS shows a great variety of morphosedimentary features, either erosional, depositional or mixed. Lonardi and Ewing (1971), although describing terraces building the slope, stressed the strong dissection by turbidity currents producing scour and erosion (even in the adjacent shelf) with high delivery and transportation of coarse sediments during sea-level lowerings. Hernández-Molina et al. (2009) described contouritic terraces, submarine canyons and steep scarps (Fig. 4.2), according to the newest interpretations that significantly improved the previous

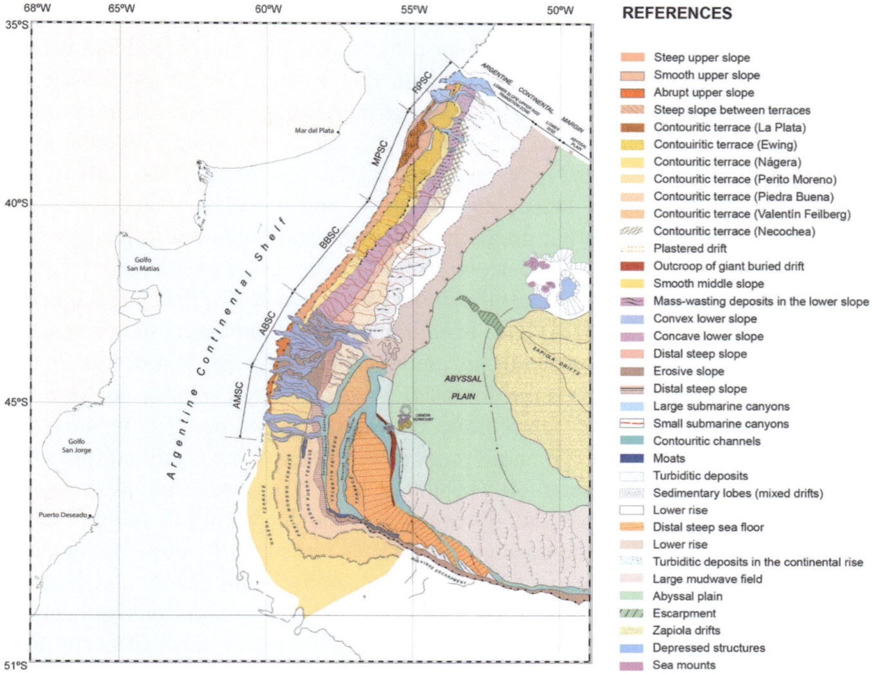

Fig. 4.2 Morphosedimentary map of the argentine slope, indicating the Contourite Depositional Systems (modified after Hernández-Molina et al. 2009). The original figure was kindly provided by F.J. Hernández-Molina

knowledge depicted (among others) by Heezen and Tharp (1961, in Heezen 1974), Ewing et al. (1964), Ewing (1965), Ewing and Ewing (1965), Lonardi and Ewing (1971), Urien and Ewing (1974), and Parker et al. (1996, 1997). Major morphosedimentary features in the southern part of the slope are four contouritic terraces (Hernández-Molina et al. 2009), named Nágera (at ~500 m water depth), Perito Moreno (~1000 m), Piedra Buena (~2100–2500 m) and Valentin Feilberg (~3500–4000 m), whereas in the northern part of the slope two contouritic terraces were originally described, Ewing Terrace (at ~1000 m water depth) and another one at 3500 m, although a later and more detailed work performed by Preu et al. (2013) defined four terraces: La Plata (T1, at ~500–600 m water depth), Ewing (T2, at 1200–1500 m), T3 (restricted to the interior of the Mar del Plata canyon at 2500 m) and Necochea Terrace (T4, 3500 m).

In the PMS five major zones can be recognized from north to south (Figs. 4.1 and 4.2): (a) Large north-Bonaerensian submarine canyons (36–38°S); (b) Contouritic terraces with low gravitational affectation (38–40°30′S); (c) Contouritic terraces dissected by a dense set of small submarine canyons (40°30′–42°30′S); (d) Large Patagonian submarine canyons (42°30′–46°S); (e) Wide Patagonian contouritic terraces (46–49°S).

(a) The northernmost part of the PMS is dominated by large submarine canyons that correspond to the Mar del Plata canyon and the Rio de la Plata canyon systems originally described by Lonardi and Ewing (1971) and Ewing and Lonardi (1971). Some detailed work on the morphology, processes and possible origin of the Mar del Plata canyon has been recently done (Violante et al. 2010a; Krastel et al. 2011; Voigt et al. 2013; Preu et al. 2010, 2013). The possible connection of these canyons with ancient fluvial systems is still a matter of debate. Although preliminary seismic and bathymetric surveys have not found any evidence of paleochannels (Ewing and Lonardi 1971; Krastel et al. 2011), some images recently studied by Urien et al. (2012, mentioned in Franco-Fraguas et al. 2014) have been interpreted as indicating the presence of channels probably representing an ancient shelf canyons connection. In this sector, dominant processes are downslope gravity-driven activity and turbidity currents, with turbiditic, debris flows and mass-wasting deposits being prevalent at the base of the slope; these processes increase to the north in direction to the Uruguayan slope (Hernández-Molina et al. 2015). Contouritic processes are highly modified here, particularly because of the decreasing in strength of the northward-flowing Antarctic-sourced currents, particularly those involved in the oceanographic system of the Brazil–Malvinas Confluence, and partially because of the pyrating of the northward-transported contourite sediments by the transverse currents inside the canyons, as documented in the Mar del Plata canyon (Preu et al. 2010; Voigt et al. 2013).
(b) This sector of the PMS is dominated by contouritic processes (Hernández-Molina et al. 2009; Violante et al. 2014b, c) because of the lack of significant submarine canyons and downslope processes. Several types of contouritic drifts and other depositional and erosive features have been described there (Violante et al. 2010b; Bozzano et al. 2011; Preu et al. 2012, 2013). The regional configuration of the Ewing and La Plata terraces (Fig. 4.2) reveals that these terraces depict here their major development and the widest extension.
(c) This area corresponds to the so-called Colorado–Negro (Lonardi and Ewing 1971) or Bahia Blanca (Hernández-Molina et al. 2009) canyon's system, which is composed of numerous (at least 20) small canyons, narrow, and most of them of relatively low morphological expression. However, judging by the high significance of depositional processes at the base of the slope (Hernández-Molina et al. 2009; Bozzano et al. 2014) (Fig. 4.2), the narrow expression of the Ewing terrace in this area, as well as the abundance of slides and mass-transport deposits in the middle slope (Costa et al. 2014), downslope processes seem to play a relevant role here. Particularly, the Bahia Blanca canyon is one of the few examples in the entire Argentine margin of a canyon entering deeply into the rise and even into the abyssal plain (Lonardi and Ewing 1971).
(d) The major submarine canyon systems develop in this area (Ameghino and Almirante Brown), which were more detailedly described by Lonardi and Ewing (1971), Rossello et al. (2005), Lastras et al. (2011) and Muñoz et al. (2012). The

particularity of these canyon systems is that in the upper section they run downslope transverse to the upper and middle slope, but in the middle–lower slope they rapidly change their orientation to the north at angles that in many cases reach 90° with respect to the upper sections, in such a way that they are derived nearly parallel to the contourlines. In the case of the Ameghino canyon system, the change in direction occurs at depths of 2000–2500 m (between 43° 20′ and 44°20′S), where a channel running downslope at a low gradient, slightly obliquous to the contourlines, collects the upper sections of the canyon valleys. At around 3500 m water depth that channel turns again downslope reaching 5000 m at 43°40′S which is considered the outlet not only of the Ameghino but also the Almirante Brown canyon systems. In the case of the Almirante Brown canyon system, a big channel nearly parallel to the contourlines runs at depths greater than ~3300 m collecting the transverse-to-the-slope upper sections of the canyons. This big channel is actually the northward continuation of a contouritic channel named Keninek (Hernández-Molina et al. 2009), and at the other end, to the north, meets the outlet of the Ameghino–Almirante Brown systems at 43°40′S. All these canyons flow toward the southwestern part of the abyssal plain in the Argentine Basin. Although tectonic implications have been considered for explaining the abrupt changes in direction of these canyon systems (Rossello et al. 2005), the most plausible interpretations (however, not denying some combination with tectonics) seem to relate the changes to the effect of strong alongslope (contouritic) currents (Hernández-Molina et al. 2009; Lastras et al. 2011; Muñoz et al. 2012), and in some cases (e.g., Lastras et al. 2011) these currents could have originated indentations that gave later origin to the upper section of the canyons by retrogressive upslope erosion. As early in 1971, Lonardi and Ewing already hypothesized that strong currents parallel to the isobaths could have exerted influence in the orientation of the middle part of the canyons.

(e) In the southernmost region of the PMS, the dominant morphosedimentary features are extensive and wide contouritic drifts that shape terraces at increasing depths, named Nagera, Perito Moreno, Piedrabuena and Valentin Feilberg (Hernández-Molina et al. 2009). Submarine canyons in this area (that correspond to the Patagonia or Deseado canyon systems described by Lonardi and Ewing 1971) are narrow and small, and in the southernmost region some of them converge into the Malvinas canyon that flows toward the Malvinas Escarpment.

The TMS is a particular feature composed of an extensive plateau that has depths at the level of the shelf and upper slope (scarcely exceeding 3000 m), except at the northern (Malvinas Escarpment) and southern (North Scotia Ridge) flanks where typical high-angle slopes develop. Along the Malvinas Escarpment and in the North Scotia Ridge (orientations broadly E–W) is very steep and narrow reaching depths close to 6000 m in the first case (south of the Argentine Basin) and close to 4000 m in the second case south of Burdwood Bank. Instead, in the Malvinas Plateau it has varied configurations with extensive and low-angle surfaces, as it accommodates around Malvinas and Georgias Islands, around shallow features like Burdwood and M.Ewing Bank, and at the sides of the Malvinas Trough; in the central part of the Malvinas Plateau the slope reaches 2000–3000 m water depth, whereas in Georgias Basin it reaches 3000–4000 m. South of the southeastern tip of Tierra del Fuego, the slope (that has here a N–S orientation) is steep and cut by at least seven submarine canyons (Lonardi and Ewing 1971). In some places of the TMS, contouritic deposits develop, like north of Burdwood Bank at the head of the Malvinas Trough (Koenitz et al. 2008; Esteban 2013; Perez et al. 2015), in the deeper part of the Malvinas Trough (Cunningham et al. 2002) and in the eastern sector of the southern flank of the North Scotia Ridge (Pudsey and Howe 2002).

Sediments: although a systematic description of the continental slope sediments is still lacking, the partial and local information provided from different bibliographic sources, as well as the information contained in the database of the Argentina Hydrographic Survey (2015), enables to depict some conclusions that can be considered as good enough to draw a general picture of the sediment distribution.

In general, the Argentine slope can be considered of composed of "coarse" sediment fractions, formed by sands and silts with a significant amount of gravels. This is due to the highly energetic environment driven by the activity of strong downslope processes able of moving even medium and coarse sands as well as gravels from the shelf to the deeper regions along the submarine canyons and steep surfaces dominated by gravity transport (turbidites, mass movements, slides, debris flows). Moreover, the highly energetic alongslope contouritic processes resulting from strong currents at the boundaries between adjacent water masses are able to transport silts and sands (and also gravels in the contouritic channels) longitudinally along the terraces in the slope. Many of these processes can also be affected by nepheloid layers, which are highly significant in the Argentine margin and make the region to be one of the most important in the world ocean in terms of suspended particles concentration (Ewing et al. 1971; Biscaye and Eittreim 1977; Emery and Uchupi 1984). Frenz et al. (2004) denounced in the slope a large amount of sand, reaching 90% at depths less than 1500 m and 20–40% at greater depths even down to 4700 m. It means a surprisingly high amount of sand at depths normally associated to fine fractions. It is also well known that inside canyons and in contouritic channels, gravels' accumulations are common (Lonardi and Ewing 1971; Bozzano et al. 2011). It has been stated (Lonardi and Ewing 1971) that at times of low and intermediate sea-level positions the amount of medium grain-size sediments delivered by rivers—that rapidly reached the slope as the shoreline was located close to the present shelf break—significantly increased as a result of the lowering in base-level as well as larger fluvial scour and erosion, so producing larger (and coarser) sediment accumulation on the slope. These processes probably increased to the south (southern Patagonian margin) as glacifluvial activity was more important there and hence a larger capacity for transporting coarse sediments is expected to have occurred, as for example the sediments pathway along the Malvinas Trough that was a significant way of transferring terrigenous materials to the lower slope

(Lonardi and Ewing 1971). Haese (1997) reported a sedimentation rate in the shelf and slope during post-glacial times around 25–50 cm/kyr. In specific depocenters, like submarine canyons, a much larger sedimentation rate can be observed, as in the case of the Mar del Plata canyon with a value of 160 cm/kyr (Voigt et al. 2013) during the Holocene.

4.5 Rise

The classical definition of continental rise considers it as the feature of the continental margins genetically associated to the final deposition of gravity-driven, downslope-transported sediments through the slope, including fan deposits at the base of submarine canyons. However, in the case of the Argentine margin other processes also intervene in its development. The rise extends continuously at the foot of the slope although it can be divided into two main regions, north and south of 44°S. North of 44°S there is a typical rise at depths deeper than 3200–3500 m (Figs. 3.3 and 4.1), formed by dominant deposition of gravity-driven sediments transported across-slope through mass-transport and turbiditic processes, as well as by supply from submarine canyons. In this region two sub-zones are distinguished, one north of 40°S where almost pure gravity processes occur (which progressively increase its influence to the north, Hernández-Molina et al. 2009, 2015; Violante et al. 2010b). Instead, between 40 and 44°S a coalescence of gravity-driven sedimentary lobes at the base of the dense net of submarine canyons develops, being the lobes deflected to the northeast by influence of south-to-north-circulating alongslope currents that modified the original turbiditic lobes by contouritic processes; in this way, the resulting features are recognized as "mixed drifts" (Hernández-Molina et al. 2009). At 44°S there is a large collecting area of downslope sedimentary fluxes coming from Ameghino and Almirante Brown submarine canyon systems.

South of 44°S the rise has different characteristics. There, alongslope contouritic processes largely dominate over downslope processes; therefore, a "rise" in the classical sense of its definition does not exists in that place. Instead, a set of contouritic terraces form the base of the slope up to the depth of the deep ocean basin (Fig. 4.2) (Hernández-Molina et al. 2009).

Sediments: although the sedimentological information of the rise is very scarce, the little available data indicate the presence of fine-grained sediments, mostly silt and clay, although with a sometimes significant sandy content particularly in turbiditic lobes and in sandy deposits located at deeper regions. Ewing and Lonardi (1971) show a very large body of what they call "thick coarse turbidites" (presumably sandy) at depths exceeding 5000 m, which were recently sampled in the region between 39 and 41°S and analyzed (Bozzano et al. 2014), demonstrating that medium-to-coarse sands dominate there associated to highly energetic downslope processes related to the Bahia Blanca submarine canyon system, with a subsequent reworking by strong deep alongslope contouritic currents.

4.6 Submarine Canyons

Although these features have been already commented when discussing the continental slope as they are relevant elements in its modeling and configuration, it is worth mentioning that they are highly significant in the ACM, although they have not been studied deeply yet. Lonardi and Ewing (1971) mapped in detail the different canyon systems crossing the slope, being only two of them extended toward the shelf break. The lack of connection with the shelf enables to consider that, in general, canyon systems should not be genetically related to fluvial influence during lowstands. Rossello et al. (2005) proposed a tectonic origin for the Ameghino and Almirante Brown canyons systems. Later on, Lastras et al. (2011) proposed a combination of gravity-driven processes and retrogressive incision for the formation of the same canyon systems. In the Mar del Plata canyon, in the northern region of the margin, no connection with the shelf has been found with seismic–bathymetric surveys, what conduced to the hypothesis of a combination of downslope processes and unstabilities with associated retrogressive evolution—possibly partially influenced by tectonism—as a main cause for the canyon formation and development (Krastel et al. 2011; Voigt et al. 2013). However, recent information provided by Urien et al. (2012, mentioned by Franco-Fraguas et al. 2014)—for the offshore region of the Rio de la Plata—alerts about the convenience of revisiting some of the concepts related to the canyons' formation as some indicators of connection between them and paleochannels preserved in the adjacent shelf have been denounced. The significance of submarine canyons is not only because of its importance as effective pathways for sediments being transferred across the margin transporting even coarse sand to the deepest environments (Ewing and Lonardi 1971; Bozzano et al. 2014) but also, as explained above, because of its influence in alongslope sedimentary processes, as recently demonstrated for the Mar del Plata canyon by Preu et al. (2010) and Voigt et al. (2013), in this last case documenting that the deposition of 7 m of Holocene sediments inside the canyon is the result of piracy, by gravity-dominated processes, of suspended materials transported alongslope by the Antarctic Intermediate Water.

4.7 Argentine Basin

The Argentine basin is the deepest oceanic environment genetically associated (although is not part of it) to the ACM, which reaches a maximum depth of $\sim$6200 m in the so-called Argentine Abyssal Plain at the southwestern part of the basin (Figs. 3.5 and 4.1). The basin plays a significant role as receiver of large amount of sediments that bypass the slope–rise system and store terrigenous materials at the "end" of the South American Source-to-Sink cycle. The basin, which covers an area of around 200,000 km^2, is bounded, apart from the Argentine margin to the west, by the Rio Grande Rise to the north, the Malvinas

Plateau/Malvinas Ridge to the south, and the Mid-Atlantic Ridge to the east. The major morphosedimentary characteristic of the basin is the development of extensive sediment drift complexes of muddy composition ("mudwaves," Lonardi and Ewing 1971; Flood et al. 1993; Manley and Flood 1993; von Lom-Keil et al. 2002) reaching thicknesses exceeding 100 m, which have a highly dynamic behavior as they migrate surrounding the basin in response to the circulation of the Antarctic Deep Water that produces a gyre in the basin before continuing to the north and pass across the Vema Channel in direction to the Brazil Basin. Most of the muddy sediments that compose the mudwaves are considered to come from the finest sediments fractions that are not retained in the shelf, slope and rise (the Argentine Margin) due to the high sediment dynamics in these environments. Three major mudwaves fields (Zapiola, Argyro and Ewing) form the relief of the basin. Regarding sedimentation rates in the abyssal plains, Stevenson and Cheng (1969) reported values between 0.5 and 5 cm/kyr.

References

Argentina Hydrographic Survey (2015). Cores and bottom samples database

Biscaye PE, Eittreim S (1977) Suspended particulate loads and transports in the nepheloid layer of the abyssal Atlantic Ocean. Mar Geol 23:155–172

Bozzano G, Violante RA, Cerredo ME (2011) Middle slope contourite deposits and associated sedimentary facies of NE Argentina. Geo-Mar Lett 31:495–507

Bozzano G, Cavallotto JL, Violante RA (2014) Los sedimentos arenosos de la Cuenca Argentina en su sector Occidental, entre Bahia Blanca y Mar del Plata. XIX Argentina geological congress (June 2014), Abstracts volume S12-8. Asoc Geol Argentina, ISBN: 978-987-22403-5-6

Campos EJD, Piola AR, Matano RP, Miller JL (2008) PLATA: a synoptic characterization of the Southwest Atlantic shelf under influence of the Plata River and Patos Lagoon outflows. Cont Shelf Res 28:1551–1555

Cavallotto JL, Violante RA, Hernández-Molina FJ (2011) Geological aspects and evolution of the Patagonian continental margin. Biol J Linn Soc 103:346–362

Cione AL, Tonni EP, Dondas A (2005) A mastodon (Mammalia, Gomphotheriidae) from the Argentinian Continental Shelf. Neues Jb Geol Palaontol Monat 10:614–630

Codignotto JO (1987) Cuaternario marino entre Tierra del Fuego y Buenos Aires. Rev Asoc Geoló Argentina 42(1–2):208–212

Codignotto JO, Kokot RR, Marcomini SC (1992) Neotectonism and sea-level changes in the coastal zone of Argentina. J Coastal Res 8(1):125–133

Codignotto JO, Kokot RR, Marcomini SC (1993) Desplazamientos verticales y horizontales de la costa argentina en el Holoceno. Rev Asoc Geol Argentina 48:125–132

Conti LA (2004) Geomorfologia da plataforma continental do Estado de São Paulo. Thesis, University of São Paulo, Brazil, p 179

Coronato AMJ, Salemme M, Rabassa J (1999) Paleoenvironmental conditions during the early peopling of Southernmost South America (Late glacial-early Holocene, 14–8 ka BP). Quatern Int 53(54):77–92

Correa ICS (1996) Les variations du niveau de la mer durant les dreniers 17.500 ans BP: l'exemple de la plate-forme continentale du Rio Grande do Sul—Brésil. Mar Geol 130:163–178

Costa IP, Violante RA, Paterlini CM, Schwenk T., Bozzano G. (2014). Deslizamientos submarinos en el talud continental adyacente al Sur bonaerense. XIX Argentina geological congress, Córdoba, Argentina. Abstracts volume S18-4. Asoc Geol Argentina, ISBN: 978-987-22403-5-6

Cunningham AP, Howe JA, Barker PF (2002) Contourite sedimentation in the Falkland Trough, western South Atlantic. In: Stow DAV, Pudsey CJ, Howe JA, Faugères J-C, Viana AR (eds) Deep-water contourite systems: modern drifts and ancient series, seismic and sedimentary characteristics. The Geological Society of London, Memoir, No 22, pp 337–352

Depetris PJ, Griffin JJ (1968) Suspended load in the Río de la Plata drainage basin. Sedimentology 11:53–60

Depetris PJ, Paolini JE (1991) Biogeochemical aspects of South American rivers: the Paraná and the Orinoco. In: Degens ET, Kempe S, Richey JE (eds) Biogeochemistry of major world rivers, SCOPE, vol 42. J Wiley & Sons, Chichester, pp 105–125

Emery KO, Uchupi E (1984) The geology of the Atlantic ocean. Springer Verlag, New York Inc, p 1050

Esteban F (2013) Estudio Geofísico-Geológico del subsuelo del segmento noroccidental de la Dorsal Norte de Scotia. Tesis Doctoral Universidad de Buenos Aires, Departamento de Ciencias Geológicas, Argentina, p 303

Ewing M (1965) The sediments of the Argentine Basin. Q J Astron Soc 6:10–27

Ewing M, Ewing J (1965) The sediments of the Argentine basin. In: Symposium oceanography Western South Atlantic, vol 37. Ann Academia Brasileira Ciencias, Suppl 31–61

Ewing M, Lonardi A (1971) Sediment transport and distribution in the Argentine Basin. 5: sedimentary structure of the Argentine margin, basin and related provinces. In: Ahrens LH, Press F, Runkorn SK, Urey HC (eds) Physics and chemistry of the earth, vol 8. Pergamon Press, London pp 125–249

Ewing M, Ludwig WJ, Ewing J (1964) Sediment distribution in the oceans: the Argentine basin. J Geophys Res 69:2003–2032

Ewing M, Eittreim SL, Ewing JI, Le Pichon X (1971) Sediment transport and distribution in the Argentine Basin. 3: Nepheloid layer and processes of sedimentation. In: Ahrens LH, Press F, Runkorn SK, Urey HC (eds) Physics and chemistry of the earth, vol 8. Pergamon Press, London, pp 49–78

Feruglio E (1950) Descripción geológica de la Patagonia. Dirección General de Yacimientos Petrolíferos Fiscales 3, Buenos Aires

Flood RD, Shor AN, Manley PL (1993) Morphology of abyssal mud waves at Project MUDWAVES site in the Argentine Basin. Deep-Sea Res II 40:859–888

Franco-Fraguas P, Burone L, Mahiques MM, Ortega L, Uriene C, Muñoz A, López G, Marin Y, Carranza A, Lahuerta N, de Mello C (2014) Hydrodynamic and geomorphological controls on surface sedimentation at the subtropical shelf front/Brazil-Malvinas confluence transition off Uruguay (Southwestern Atlantic continental margin). Mar Geol 349:24–36

Frenz M, Höppner R, Stuut J-BW, Wagner T, Henrich R (2004) Surface sediment bulk geochemistry and grain-size composition related to the oceanic circulation along the South American Continental Margin in Southwest Atlantic. In: Wefer G, Mulitza S, Ratmeyer V (eds) The South Atlantic in the late quaternary. Springer-Verlag, pp 347–373

Gaiero DM, Probst JL, Depetris PJ, Bidart SM, Leleyter L (2003) Iron and other transition metals in Patagopnian riverborn and windborne materials: geochemical control and transport to the South Atlantic Ocean. Geochim Cosmochim 67:3603–3623

Gaiero DM, Depetris PJ, Probst J-L, Bidart S, Leleyter L (2004) The signature of river- and wind-borne materials exported from Patagonia to the southern latitudes: a view from REEs and implications for paleoclimatic studies. Earth Planet Sci 219:357–376

Groeber P (1948) Las plataformas submarinas y su edad. Rev Cienc Invest 6:224–231

Haese RR (1997). Beschreibung und Quantifizierung fr¨uhdiagenetischer Reaktionen des Eisens in Sedimenten des Südatlantiks. Berichte, Fachbereich Geowissenschaften, Universität Bremen, 99: 118

Heezen BC (1974) Atlantic-type continental margins. In: Drake Ch (ed) Burk C. The Geology of Continental Margins, Springer-Verlag, New York, pp 13–24

Heezen BC, Menard HW (1966) Topography of the deep-sea floor. In: Hill MN (ed) The sea. Intersciences Publ, 2nd Edn, J Wiley & sons: 233–280

Heezen BC, Tharp M (1961) Physiographic diagram of the South Atlantic. Bull Geol Soc Am

Hernández-Molina FJ, Paterlini CM, Violante RA, Marshall P, de Isasi M, Somoza L, Rebesco M (2009) A contourite depositional system on the Argentine slope: an exceptional record of the influence of Antarctic water masses. Geology 37(6):507–510

Hernández-Molina FJ, Soto M, Piola AR, Tomasini J, Preu B, Thompson P, Badalini G, Creaser A, Violante RA, Morales E, Paterlini CM, de Santa Ana H (2015) A contourite depositional system along the Uruguayan continental margin: sedimentary, oceanographic and paleoceanographic implications. Mar Geol. doi:10.1016/j.margeo.2015.10.008

Isla FI (1989) Holocene sea-level fluctuation in the southern hemisphere. Quatern Sci Rev 8:359–368

Isla FI (2011) Submerged woods of Upper Pleistocene age at San Matías Gulf: a sea level curve for Northern Patagonia. In: 13° Congreso de la Asociación Brasileña de estudios del Cuaternario ABEQUA, Anales CD, Buzios, p 5

Isla FI, Bujalesky GG (2008) Coastal Geology and morphology of Patagonia and the Fuegian Archipielago. In: Rabassa J (ed) The Late Cenozoic of Patagonia and Tierra del Fuego. Dev Quat Sci 11: 227–240

Isla FI, Madirolas A (2009) Submerged Pleistocene highstand at the inner shelf of the Rio de la Plata: the assymetry of the La Plata (Pez Limón) bank. Thalassas 25(2):21–26

Isla FI, Schnack E (1995) Submerged moraines offshore northern Tierra del Fuego, Argentina. Quat S Am Antarct Peninsula 9:205–222

Kennett J (1982) Mar Geol, Prentice Hall, 813

Koenitz D, White N, McCave IN, Hobbs R (2008) Internal Structure of a contourite drift generated by the Antarctic circumpolar current. Geochem Geophys Geosyst 9(6):Q06012. doi:10.1029/2007GC001799

Krastel S, Wefer G, Hanebuth TJJ, Antobreh AA, Freudenthal T, Preu B, Schwenk T, Strasser M, Violante RA, Winkelmann D, M78-3 shipboard scientific party (2011) Sediment dynamics and geohazards offshore Uruguay and Northern Argentina: First results from the multi-disciplinary Meteor-Cruise M78-3. GeoMarine Lett 31:271–283

Kuenen H (1950). Marine geology. J.Wiley & Sons, p 568

Lastras G, Acosta J, Muñoz A, Canals M (2011) Submarine canyon formation and evolution in the Argentina Continental Margin between 44°30′S and 48°S. Geomorphology 128:116–136

Lonardi A, Ewing M (1971) Sediment transport and distribution in the Argentine Basin. In: Press F, Runkorn SK, Urey HC (eds) Physics and Chemistry of the Earth, Pergamon Press, London vol 8, pp 253–264

Mahiques MM, de Mello e Sousa SH, Furtado VV, Gonsalez Tessler M, de Lima Toledo FA, Burone L, Lopes Figueira RC, Klein DA, Martins CC, Alves DPV (2010) The southern Brazilian shelf: general characteristics, Quaternary evolution and sediment distribution. Braz J Oceanogr 58: 25–34

Manley PL, Flood RD (1993) Paleo£ow history determined from mudwave migration; Argentine Basin. In: Manley PL, Flood RD (eds) Project mudwaves. Deep-Sea Res II 40:1033–1055

Melo WD, Schillizzi R, Perillo GME, Piccolo MC (2003) Influencia del área continental pampeana en la evolución morfológica del estuario de Bahía Blanca. AAS, Rev Asoc Argentina Sedimentología 10(1):37–50

Melo WD, Perillo GME, Perillo MM, Schilizzi R, Piccolo MC (2013) Late Pleistocene–Holocene deltas in southern Buenos Aires province, Argentina. Deltas: landforms, ecosystems and human activities. In: Proceedings of HP1, IAHS-IAPSO-IASPEI Assembly, Gothenburg, Sweden. IAHS publication, vol 358, pp 187–195

Mouzo F (2005) Límites de las glaciaciones Plio-Pleistocenas en la plataforma continental al noreste de la Tierra del Fuego. XVI Congreso Geol Argentino, Actas 3:787–792

Mouzo F (2014) Edad del golfo San Matías, Plataforma Continental Argentina y la estratigrafía en el Gran Bajo de San Antonio, Provincia de Río Negro. Rev Asoc Geol Argentina 71(1):125–138

Mouzo FH, Garza ML, Izquierdo J, Zibecchi RO (1978) Rasgos de la geología submarina del golfo Nuevo (Chubut). Acta Oceanogr Argentina 2:69–91

Muñoz A, Cristobo J, Rios P, Druet M, Polonio V, Uchupi E, Acosta J, Atlantis Group (2012) Sediment drifts and cold-water coral reefs in the Patagonian upper and middle continental slope. Mar Pet Geol 36(1):70–82

Osterrieth M, Violante RA, Borrelli N (2008). Silicofitolitos en sedimentos de testigos marinos de la región norte de la Plataforma Continental Argentina. In: Korstanje MA, Babot MP (eds) Matices Interdisciplinarios en Estudios Fitolíticos y de otros Microfósiles. Br Archaeol Res Ser 1870(10): 109–117

Parker G, Violante RA, Paterlini CM (1996) Fisiografía de la Plataforma Continental. In: Ramos V, Turic M (eds) Geología y Recursos Naturales de la Plataforma Continental Argentina. Relatorio XIII Congreso Geológico Argentino. Asociación Geológica Argentina, Buenos Aires, pp 1–16

Parker G, Paterlini CM, Violante RA (1997) El fondo marino. In: Boschi E (ed) El Mar argentino y sus Recursos Pesqueros. INIDEP, Mar del Plata vol 1, pp 65–87

Parker G, Paterlini CM, Violante RA, Costa IP, Marcolini S, Cavallotto JL (1999) Descripción Geológica de la Terraza Rioplatense (Plataforma Interior del Noreste Bonaerense). Servicio Geol Min Argentino, Bol (273): 98

Parker G, Violante RA, Paterlini CM, Marcolini S, Costa IP, Cavallotto JL (2008) Las secuencias sismoestratigráficas del Plioceno-Cuaternario en la Plataforma Submarina adyacente al litoral del este bonaerense. Lat Am J Sedimentology Basin Anal 15:105–124

Paterlini CM, Mouzo F (2013) Exploración sísmica del Golfo Norpatagónico San José, Plataforma Continental Argentina. Provincia del Chubut. Geoacta 38(1):81–94

Pedoja K, Regard V, Husson L, Martinod J, Guillaume B, Fucks E, Iglesias M, Weill P (2011) Uplift of quaternary shorelines in eastern Patagonia: Darwin revisited. Geomorphology 127:121–142

Pérez LF, Hernández-Molina FJ, Esteban FD, Tassone A, Piola AR, Maldonado A, Preu B, Violante RA, Lodolo E (2015) Geo-Mar Lett 35(4):271–288

Perillo GME, Kostadinoff J (2005) Margen Continental de la Provincia de Buenos Aires. Relatorio Geología de la Provincia de Buenos Aires. XVI Congreso Geológico Argentino, La Plata, pp 277–292

Pierce JW, Siegel FR (1979) Suspended particulate matter on the Southern Argentina Shelf. Mar Geol 29:73–91

Ponce JF, Rabassa J, Coronato A, Borromei AM (2011) Paleogeographic evolution of the Atlantic coast of Pampa and Patagonia since the Last Glacial Maximum to the Middle Holocene. Biol J Linn Soc 103:363–379

Preu B, Schwenk T, Hernández-Molina FJ, Strasser M, Huppertz T, Hanebuth TJJ, Violante RA, Krastel S, Henrich R, Lindhorst K, Meyer M Spieß V, M78/3 shipboard scientific party (2010) Interaction between a Contourite depositional system and the Mar del Plata Canyon off Argentina and Uruguay. Geotemas 11:139–140

Preu B, Schwenk T, Hernández-Molina FJ, Violante RA, Paterlini CM, Krastel S, Tomasini J, Spieß V (2012) Sedimentary growth pattern on the northern Argentine slope: the impact of North Atlantic deep water on southern hemisphere slope architecture. Mar Geol 329–331:113–125

Preu B, Hernández-Molina FJ, Violante RA, Piola AR, Paterlini CM, Schwenk T, Voigt I, Krastel S, Spieß V (2013) Morphosedimentary and hydrographic features of the northern Argentine margin: the interplay between erosive, depositional and gravitational processes and its conceptual implications. Deep-Sea Res 75:157–174

Pudsey CJ, Howe JA (2002) Mixed biosiliceous-terrigenous sedimentation under Antarctic Circumpolar Current, Scotia Sea. In: Stow DAV, Pudsey CJ, Howe JA, Faugères J-C, Viana AR (eds) Deep-water contourite systems: modern drifts and ancient series, seismic and sedimentary characteristics. The Geological Society of London, Memoir, No 22 pp 325–336

Rossello EA, Lagabrielle Y, Cobbold PR, Marshall P (2005) Los cañones submarinos oblicuos del talud continental argentino (40° a 45°S): evidencias de inversión tectónica andina sobre el margen pasivo atlántico?. X Simposio Nacional de Estudos Tectónicos-IV International Symposium on Tectonics, Curitiba, Brasil

Rostami K, Peltier WR, Manzini A (2000) Quaternary marine terraces, sea-level changes and uplift history of Patagonia, Argentina: comparisons with predictions of the ICE-4G (VM2) model of the global process of glacial isostatic adjustment. Quatern Sci Rev 19:1495–1525

Schellmann G, Radtke U (2000) ESR dating of stratigraphically well-constrained marine terraces along the Patagonian Atlantic coast (Argentina). Quatern Int 68–71:261–273

Shepard FP (1973). Submarine geology 3rd Edn. Harper & Row, New York, p 551

Spalletti LA, Isla FI (2003) Características y evolución del delta del Río Colorado ("Colú-Leuvú"), Provincia de Buenos Aires, República Argentina. AAS, Rev Asoc Argentina Sedimentología 10:23–37

Stevenson FJ, Cheng C-N (1969) Amino acid levels in the Argentine basin sediments: correlation with quaternary climatic changes. J Sediment Petrol 39:345–349

Urien CM, Ewing M (1974) Recent sediments and environments of Southern Brazil, Uruguay, Buenos Aires and Río Negro Continental Shelf. In: Burk C, Drake Ch (eds) The geology of continental margins, Springer-Verlag, New York, pp 157–177

Urien CM, Paterlini M, Berbeglia Y, Jimenez M (2012) La Cuenca del Plata: un tema para pensar, II Workshop Antropicosta Iberoamerica, Montevideo, Uruguay. 40

Violante RA (2005) Submerged terraces in the continental shelf of Argentina and its significance as paleo-sea level indicators: the example of the Rioplatense Terrace. 5th Annual conference Project IGCP 464 "Continental shelves during the last glacial cycle", San Petersburg, Rusia. Abstract volume: pp 97–99

Violante RA, Parker G (2000) El Holoceno en las regiones costeras y marinas del noreste de la Provincia de Buenos Aires. Rev Asoc Geol Argentina 55(4):337–351

Violante RA, Parker G (2004) The post-Last Glacial Maximum transgression in the de la Plata river and adjacent inner continental shelf, Argentina. Quat Int 114(1):167–181

Violante RA, Osterrieth M, Borrelli N (2007) Evidences of subaerial exposure of the Argentine Continental Shelf during the Last Glacial Maximum. XVII Congreso Internacional de INQUA y 1° Reunion Anual Proyecto UNESCO-IUGG—IGCP 526, Cairns, Australia. Abstracts, Vol. Quatern Int 167–168:434

Violante RA, Paterlini CM, Bozzano G, Hanebuth TJJ, Strasser M, Krastel S (2010a) Morphosedimentary characteristics and evolution of the Mar del Plata Submarine Canyon, Northern Argentina Continental Margin. 18° International Sedimentological Congress, Mendoza, Argentina, Abstracts, p 907

Violante RA, Paterlini CM, Costa IP, Hernández-Molina J, Segovia LM, Cavallotto JL, Marcolini S, Bozzano G, Laprida C, García Chapori N, Bickert T, Spiess V (2010b) Sismoestratigrafía y evolución geomorfológica del talud continental adyacente al litoral del este bonaerense, Argentina. Lat Am J Sedimentology Basin Anal 17:33–62

Violante RA, Bozzano G, Cavallotto JL, Marcolini S, Blasi A, Laprida C (2014a) El paleosistema fluvio-estuárico-deltaico del río Colorado en la plataforma continental del sur bonaerense. XIX Argentina Geological Congress, Abstracts volume S12-42. Asociación Geológica Argentina, ISBN: 978-987-22403-5-6

Violante RA, Costa IP, Cavallotto JL, Paterlini CM, Marcolini S, Bozzano G (2014b) Rasgos Morfosedimentarios, Procesos y Evolución de la Plataforma Continental Argentina desde el Último Máximo Glacial. Rev Asoc Geol Argentina 71(2): 292–310

Violante RA, Hernández-Molina FJ, Costa IP, Schwenk T (2014c) Different styles in configuration of the contouritic drifts in the northern sector of the Argentina continental margin: implications in slope stability and geohazard. 2nd Deep-water circulation congress, Ghent, Belgium. Abstracts, pp 17–18

Violante RA, Laprida C, Bressan GS, Diaz GR, Bozzano G, Grant JP, Garcia Chapori N, Cavallotto JL, Maidana NI, Cianfagna FA, Osterrieth ML, Paterlini CM, Costa IP (2014d) Registro paleoambiental del Estadio Isotópico 2 en la Plataforma Continental exterior del Sureste Bonaerense: un nuevo aporte a la evolución regional. Rev Asoc Geol Argentina 71(3): 449–455

Violante RA, Marcolini S, Cavallotto JL, Paterlini CM, Costa IP, Laprida C, Dragan, W, Watanabe S, Totah V, Rovere EI, Osterrieth ML (2014e) The Argentina continental shelf: morphology, sediments, processes and evolution since the Last Glacial Maximum. In: Chivas AR, Chiocci FL (eds) Continental shelves during last Glacioeustatic cycle: shelves of the world reviews. Final volumen Project IUGG-UNESCO-IGCP 464 "Continental Shelves during the Last Glacial Cycle". Geological Society of London, Memoirs vol 41, pp 55–68

Voigt I, Henrich R, Preu BM, Piola AR, Hanebuth TJJ, Schwenk T, Chiessi CM (2013) A submarine canyon as a climate archive—Interaction of the Antarctic Intermediate water with the Mar del Plata Canyon (Southwest Atlantic). Mar Geol 341:46–57

von Lom-Keil H, Spieß V, Hopfauf V (2002) Fine-grained sediment waves on the western flank of the Zapiola drift, Argentine Basin: evidence for variations in late quaternary bottom flow activity. Mar Geol 192:239–258

Chapter 5
Climate and Oceanographic Background

Abstract The regional climatic and oceanographic characteristics that regulate the climate-ocean system in the region are explained. The climatic system has different features depending on regional and local forcings, such as the Intertropical Convergence Zone, the South American Monsoon System, the Southern Hemisphere Westerlies, and the South Atlantic Convergence Zone. The hydrographic structure is dominated by different water-masses of Antarctic and Tropical origin, related to the major circulation system of the Western South Atlantic linked to the Atlantic Meridional Overturning Circulation. The main water-masses constituting the hydrographic structure are the Tropical Water-South Atlantic Central Water, the Antarctic Intermediate Water, the North Atlantic Deep Water and the Antarctic Bottom Water, being most of the Antarctic-sourced water-masses genetically related to the circulation of the Antarctic Circumpolar Current.

Keywords Climate-oceanography coupling · South America climate · Oceanography · Ocean Meridional Overturning Circulation · Water-masses · Brazil–Malvinas Confluence

Climate and oceanographic processes, which are strongly coupled in the atmosphere–sea system, play a relevant role in the particular region of the Western South Atlantic. For this reason, some introductory background is given in order to explain the atmospheric and oceanic forcing factors involved in the evolution, development and resulting sedimentary and biological constitution of the Argentina Continental Margin.

5.1 Regional Climate Background

South America is a particular region of the globe as it is the largest landmass in the ocean-dominated Southern Hemisphere (Clapperton 1993). The significant meridional (north-south) extension of the continent encompassing several climatic zones

R.A. Violante et al., *The Argentina Continental Margin*,
SpringerBriefs in Earth System Sciences, DOI 10.1007/978-3-319-04196-4_5

from the tropics (12°N) to mid (subpolar) latitudes (55°S), as well as the southward narrowing of the land with prominent topographic differences in relatively short distances, are major attributes that, added to the great oceanic influence, produce particular climatic and even oceanographic conditions, with diverse patterns of weather and climate. A relevant conditioning factor is the Andes Cordillera, a morphological and climatic barrier which extends all along the western coast of the continent.

South America climate is affected by remote, regional and local forcings (Solman 2013). As a consequence, three major climate zones can be distinguished: tropical South America, subtropical South America, and austral South America.

Atmospheric circulation and climate in all three zones is highly modulated and constrained by the orography of the Andes, the shape of the continent, and the interactions with the underlying land-surface, vegetation and soil moisture (Flantua et al. 2015). At the mid-latitudes south to 36°S, the Andes not only act as a climatic barrier ("climatic wall") with moist conditions to the west and dry conditions to the east, but also foment tropical-extratropical interactions, especially along their eastern side (Garreaud et al. 2009). The variability of the South America climate from intraseasonal to interdecadal timescales results from the superposition of several remote large-scale phenomena: El Niño-Southern Oscillation (ENSO), the control exerted by anomalous sea-surface temperature (SST) over subtropical Atlantic and Pacific Oceans, SST anomalies over the tropical North Atlantic Ocean which affect climate variability over the Amazonia and Northeastern Brazil, and high-latitude forcings such as by the Antarctic Oscillation (AAO) and the North Atlantic Oscillation (NAO) (Solman 2013).

5.1.1 Intertropical Convergence Zone

Tropical and subtropical features in South America include an east–west oriented narrow band over the tropical oceans called the Intertropical Convergence Zone (ITCZ) (Fig. 5.1). The ITCZ corresponds to the belt of minimum pressure and intense low-level convergence of the tradewinds over the equatorial Pacific and Atlantic oceans foster precipitation maxima. The seasonal migration of the ITCZ largely controls climate in the northern part of the continent. Over the Atlantic sector, the ITCZ reaches the Equator, producing the rainy season of northeast Brazil. In contrast to the abundant precipitation near the ITCZ, rainfall is almost absent over broad areas of the subtropical Atlantic Ocean due to the large-scale mid-tropospheric subsidence (Garreaud et al. 2009). The subsidence also maintains a semipermanent high-pressure cell at about 30°S. The subsidence over the SW Atlantic is most intense in the austral winter but encompass a larger meridional extent in the austral summer (Dima and Wallace 2003).

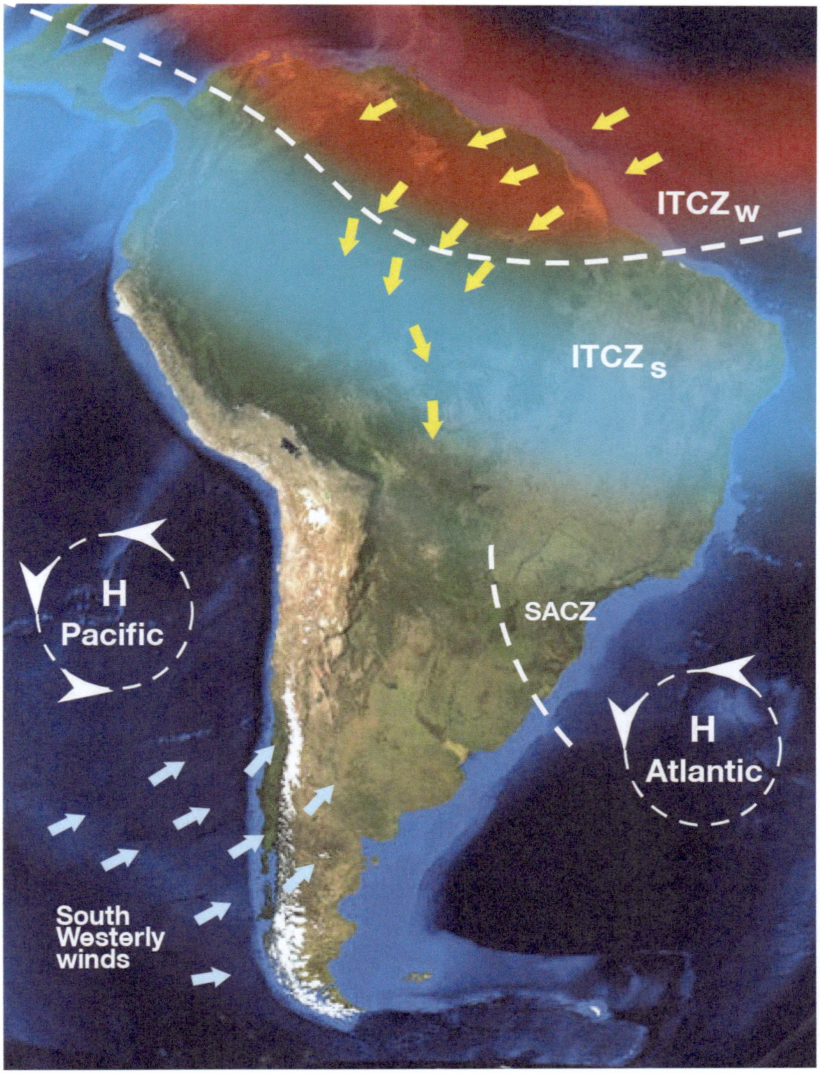

Fig. 5.1 Basic climatic map of South America. *H* Subtropical highs. *ITCZ w* and *s* winter and summer (respectively) Intertropical Convergence Zone. *SACZ* South Atlantic Convergence Zone

5.1.2 South American Monsoon System

Consistent with the low thermal inertia of the land, tropical and subtropical rainfall over the continent experiences a pronounced seasonal cycle (Horel et al. 1989; Fu et al. 1998; Marengo et al. 2001). During the austral winter, the maximum rainfall in the continent is located to the north of the Equator, almost in line with the ITCZ.

Meanwhile, the central part of the continent experiences its dry season. By the end of October there is a rapid southward shift of the convection, so that during the austral summer a broad area of heavy precipitation extends from the southern of the Amazon Basin to northern Argentina. The large-scale circulation at upper levels is characterized by a high-pressure center over the Altiplano. At low levels, the semipermanent subtropical highs over the Atlantic and Pacific Oceans dominate the large-scale features (Fig. 5.1). Easterly flow from the Atlantic Ocean is channeled southwards by the Andes mountains into the Chaco low (low-pressure zone), which represents the main source of moisture over southern Brazil and the subtropical plains of Argentina. During the austral fall, the precipitation maximum returns gradually to the north. Such migration has led many to describe the climate of the central part of South America as Monsoon-like (Zhou and Lau 1998; Vera et al. 2006), although throughout the year the trade winds over the Atlantic blow toward the continent (albeit with different angles) where the pressure is lower than over the ocean (Garreaud et al. 2009). The South American Monsoon System is one of the major components of the continental warm season precipitation regime over tropical and subtropical latitudes, while over the southern part of the continent synoptic activity is dominated by the presence of the Pacific storm track and its interaction with the Andes.

5.1.3 Southern Hemisphere Westerlies

The belt of Southern Hemisphere Westerlies (Fig. 5.1) is largely symmetric and has a rather modest annual cycle due to the absence of significant land masses south of 35°S (Nakamura and Shimpo 2004). South of 40°S, low-level westerly flow prevails year-round over continental South America and the adjacent oceans. Over southern South America, the westerlies are strongest during the austral summer, peaking between 45°S and 55°S. During the austral winter, the low-level westerlies expand equatorward and the band moves into subtropical latitudes peaking at about 30°S, and weakens at $\sim$50°S.

5.1.4 South Atlantic Convergence Zone

Precipitation over extratropical South America exhibits a marked zonal asymmetry. The western slope of the Andes produces orographic precipitation leading to a local maximum 2–3 times larger than the corresponding oceanic values. In contrast, forced subsidence over the eastern side of the Andes produces very dry conditions in Argentina's Patagonia. Significant frontal rainfall reappears near the Atlantic coast. Over the Atlantic, an extratropical diagonal band of precipitation maxima is known as the South Atlantic Convergence Zone (SACZ, Fig. 5.1) (Kodama 1992; Liebmann et al. 1999; Carvalho et al. 2004). The SACZ is a region of intense convection and represents an important feature of the summer circulation in South America. This band is oriented northwest to southeast from a region of intense convection over the Amazon Basin in northern South America and is projected into the South Atlantic Ocean crossing the Brazilian coast at about 20°S. The combined action of Amazonian latent heat source and the Andean topography is the mechanism for the generation of the SACZ (Figueroa et al. 1995). The SACZ is more intense during summer when it is connected with the area of convection over the central part of the continent, producing episodes of intense rainfall over much of southeastern South America (Liebmann et al. 1999). Periods of enhanced SACZ activity are associated with an excess of precipitation in its core and the southern coast of Brazil, and drier than normal conditions in northern Argentina, Paraguay and Uruguay. Roughly symmetric but opposite conditions prevail during the weak SACZ periods. At interannual timescales SACZ variability is associated with an anomalous upper tropospheric, large-scale stationary eddy in the lee of the Andes (Robertson and Mechoso 2000), which in turn is significantly correlated with SST anomalies over the South Atlantic.

In the Argentine coastal zone, the precipitation annual cycles are different in the north and the south having different controls (Fig. 5.1). In the north, the rainfall maximum coincides with the increase on the cyclogenetic activity from autumn to winter, while in the south the rainfall distribution is almost uniform throughout the year. Larger positive correlation between turbulent fluxes and precipitation annual cycles are obtained, which suggest the contribution of a local source of moisture for the precipitation (Reboita et al. 2010).

5.2 Oceanographic Background

From the oceanographic point of view, the Western South Atlantic Ocean is a key region in the global oceanic circulation system, and a unique place in the world ocean where equatorial and south polar water-masses actively interact with a net transport of meridional heat between the Southern Pole and the Equator (Bryden and Imawaki 2001; Talley 2003). In the following sections the main oceanographic components that are part of that system in the Argentina Continental Margin are described.

5.2.1 The Western South Atlantic and the Meridional Overturning Circulation

The Meridional Overturning Circulation (MOC) is a global-scale ocean circulation system driven by the Equator-to-Pole surface density differences of seawater (Figs. 2.4 and 5.2). As a major part of the general circulation, the MOC supplies deep water to all the ocean basins (Schloesser et al. 2012). Its deep southward flow of cold and salty North Atlantic Deep Water (NADW) along the eastern coast of South

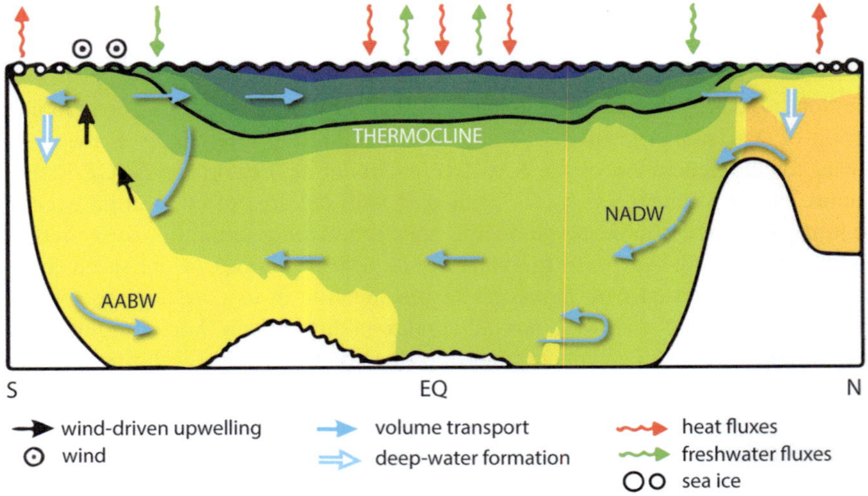

Fig. 5.2 Schematic disposition of main water-masses in the Atlantic Ocean. *AABW* Antarctic Bottom Water. *NADW* North Atlantic Deep Water. Modified after Kuhlbrodt et al. (2007)

America, compensated by a warm northward return flow at surface, central and intermediate waters, makes of the Meridional Overturning Circulation one of the primary pathways by which the ocean transports heat poleward (e.g., Toggweiler and Samuels 1995; Munk and Wunsch 1998). In the Atlantic Ocean it is often referred to as the Atlantic Meridional Overturning Circulation (AMOC). This system consists of four main processes (Fig. 5.2): (a) the surface transport of water toward high latitudes; (b) the deep transport of water in the opposite way; (c) the cold nutrient-rich waters transport from depth to near surface due to upwelling processes; and (d) the deep water formation in regions where waters become denser and cooler and as a result of this they sink. This results in a circulation system of two overturning cells: the deep one with NADW and an abyssal one with Antarctic Bottom Water (AABW) (Fig. 5.2) (Kuhlbrodt et al. 2007). Nowadays, two distinct mechanisms for driving the AMOC are being debated. One is the traditional thermohaline driving mechanism, originally proposed by Sandström (1916) and Jeffreys (1925), where the driver is the (diapycnal) mixing that transports heat from the surface to the deep water-masses (Fig. 5.2) (Munk and Wunsch 1998). In this view, internal waves are generated by the action of winds and tides. These waves dissipate into small-scale motion what causes turbulent mixing of heat that lightens the deep water-masses, making them to rise in low latitudes. The resulting waters are advected poleward into the North Atlantic where they become denser by the atmospheric cooling and salt rejection during sea-ice growth, and consequently they sink (Fig. 5.2) (Kuhlbrodt et al. 2007). The other mechanism is the wind-driven upwelling (Toggweiler and Samuels 1993, 1995, 1998). As an alternative, those authors proposed that most of the oceanic upwelling is wind-driven and occurs in the Southern Ocean (Drake Passage Effect). The strong westerly circumpolar winds induce a northward transport of surface waters, and as a result of the Ekman transport, an upwelling from depth is induced (Fig. 5.2). Within both scenarios, the South Atlantic represents an essential role in the AMOC, not only because its obvious connection with the North Atlantic, but also because through the Antarctic Circumpolar Current (ACC) it connects the Pacific and Indian Oceans, representing the most important global-connecting oceanic circulation (Garrison 2008). Despite of this, there is still great uncertainty on the absolute magnitude of the South Atlantic inter-ocean fluxes. Far from being just a passive conduit for the transit of the water-masses, it actively influences these through air–sea interactions, mixing, as well as through subduction and advection processes (Garzoli and Matano 2011).

5.2.2 Western South Atlantic Main Currents

The South Atlantic is dominated by a system of oceanographic fronts that results in three zones of relatively uniform water properties: the Subtropical Front Zone, the Subantarctic Front Zone, and the Antarctic Polar Front Zone (Fig. 5.3). The Subtropical Front represents the southern boundary of the anticyclonic Subtropical Gyre and separates the gyre circulation from the Subtropical Zone (Peterson and Stramma 1991). The eastern boundary current of the Subtropical Gyre, characterized by strong upwelling, is the Benguela Current (e.g., Lutjeharms and Meeuwis 1987; Lutjeharms and Valentine 1987; Shannon et al. 1990). The Brazil Current forms the western boundary current of the gyre, transporting tropical warm and salty waters toward the south (Fig. 5.3) (Piola and Matano 2001). Warm salty nutrient-poor waters characterize the Subtropical Zone, which southern boundary is determined by the Subantarctic Front, represented by an abrupt decline in salinity and temperature of surface waters (Fig. 5.3). This defines the Subantarctic Zone. Finally, it is the Antarctic Polar Zone, which is delimited to the north by the Antarctic Polar Front and to the south by the Antarctic continent (Fig. 5.3). This zone is characterized by the dominance of waters with very high nutrient content and sea-surface temperature lower than 10 °C, as well as by the seasonal formation of sea-ice. Here, the sinking along the Antarctic Polar Front, being the most extensive intermediate depth water-mass in the World Ocean (Gordon 1981), forms the Antarctic Intermediate Water (AAIW).

Offshore, the western South Atlantic presents a highly dynamic frontal zone in surface: the Brazil-Malvinas Confluence (BMC), bounded by two surface western boundary currents, the already mentioned Brazil Current and the Malvinas Current (Fig. 5.3) (Gordon 1981; Peterson and Stramma 1991; Stramma and England 1999). The dynamic balance between the opposing transports of these two currents determines the Confluence location (Matano 1993). The Brazil Current originates in the bifurcation of the South Equatorial Current at $\sim$15°S. It carries warm and salty waters along the continental slope of South America toward the south. Between 35 and 39°S it encounters the Malvinas Current that carries cold and well-oxygenated waters of subantarctic origin toward the Equator (Fig. 5.3) (Piola and Gordon 1989).

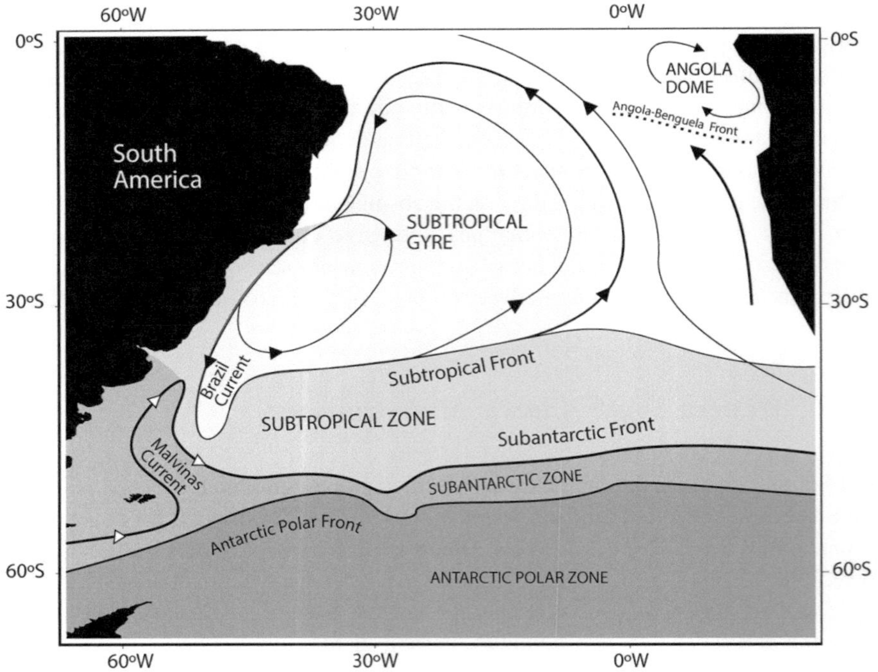

Fig. 5.3 Main surface currents and oceanographic fronts in the South Atlantic [modified after García Chapori et al. 2015, who reproduced the original figure from Peterson and Stramma (1991) and Stramma and England (1999)]

This current represents the septentrional branch of the Antarctic Circumpolar Current (ACC), flowing northward along the Argentine margin. The Malvinas Current is strongly influenced by the south westerly winds, tidal range and the low-salinity discharges from the continent (Palma et al. 2008; Matano et al. 2010; Piola et al. 2010). As a result, the primary productivity of the region is exceptionally high. In the Patagonian margin, the interaction between this downwelling current and the continental slope produces what it is known as the "Shelf-break Upwelling of Patagonia" (Matano and Palma 2008; Matano et al. 2010) and, in consequence, the surface waters enrich in nutrients producing a permanent high-chlorophyll zone along the flow of the current (Gappa 2000; Matano and Palma 2008; Matano et al. 2010). The encounter of the Brazil and Malvinas currents generates sharp horizontal and vertical gradients in temperature, salinity, density and nutrient content, mixing the main water-masses present in the region (Table 3.1): the Tropical Water (TW), the South Atlantic Central Water (SACW) and the Antarctic Intermediate Water (AAIW) (Gordon 1989; Peterson and Stramma 1991; Bianchi et al. 1993; Wilson and Rees 2000; Piola and Matano 2001). Their interaction dominates the oceanographic circulation system between ~29°S and 49°S (Peterson and Stramma 1991; Stramma and England 1999), making the western South Atlantic a natural target of several ocean-related studies (e.g.,: Boltovskoy et al. 1996; Gordon 1981; Peterson and Stramma 1991; Stramma and England 1999; Piola and Matano 2001; Henrich et al. 2003; Chiessi et al. 2007, 2014; Toledo et al. 2007, 2008; Laprida et al. 2011).

The BMC presents a substantial latitudinal drift, including meridional migration at both seasonal and interannual timescales that involve displacements of several hundred of kilometers (Olson et al. 1988; Bianchi and Garzoli 1997). It drifts $\sim 2°$ equatorward (poleward) during the austral winter (summer) (Boltovskoy 1981; Legeckis and Gordon 1982; Olson et al. 1988). However, recent studies evidenced an additional southward drift during the period between 1993 and 2008 (Goni et al. 2011; Combes and Matano 2014). Satellite data and numerical models agree in that the BMC has drifted southward at a rate of 0.39°–0.81°/decade during that period. Several studies have linked these periodic variations of the Confluence location to wind stress curl over the subtropical gyre and/or the Antarctic Circumpolar Current transport at the Drake Passage (Matano et al. 1993; Garzoli and Giulivi 1994; Wainer et al. 2000; Goni and Wainer 2001; Fetter and Matano 2008). However, recent contributions associate them with an abrupt generalized weakening of the Southern Ocean circulation produced by a weakening of the westerly wind forcing (Combes and Matano 2014).

The western South Atlantic shelf, which extends from Cape Frio (Brazil) to Burdwood Bank, is also characterized by the southward flow of warm waters in the Brazilian Margin and the northward flow of cold waters in the Argentine Margin (Piola et al. 2000; Palma et al. 2008). Near 33°S, the fresh nutrient-rich Subantarctic Shelf Water and the warm salty nutrient-poor Subtropical Shelf Water collide, resulting in what is known as the Subtropical Shelf Front (Fig. 5.3) (Piola et al. 2000, 2008). To the south of this latitude, the Malvinas Current percolates over the Patagonian shelf influencing the regional ecosystems (Piola et al. 2010); whereas to the north, the Brazil Current presents sporadic onshore intrusions of eddies and meanders as its most important contribution to the shelf dynamics (Matano et al. 2010). However, even when the Subtropical Shelf Front seems to be the northwestward shallow-water extension of the BMC, the processes controlling its latitudinal position are not the same and are still under discussion. Some authors have suggested that the cross-shelf scale of the pressure gradient enforced by the Malvinas Current and its northernmost extension play a major role in controlling the latitudinal position of the Subtropical Shelf Front (Matano et al. 2010; Palma et al. 2008). As a consequence, at surface its position shows a $\sim 5°$ northward displacement during the austral winter, but below the mixed surface layer, the Subtropical Shelf Front position in the modern system experiences no significant seasonality (Möller et al. 2008; Piola et al. 2008). The latitudinal shifts of both, the Subtropical Shelf Front and the BMC, are not confined just to the present, but they have occurred along the most recent geologic time-period. Recent contributions have added novel and direct evidence for latitudinal shifts of both, the Subtropical Shelf Front and the BMC during the Late Pleistocene and Holocene (e.g.,: Laprida et al. 2011; Bender et al. 2013; Chiessi et al. 2014; Voigt et al. 2015).

5.2.3 Western South Atlantic Hydrographic Structure

The western South Atlantic presents a highly dynamic hydrographic structure. Depths above 500 m are dominated by the southward flow of the Tropical Water (TW) and South Atlantic Central Water (SACW), and the northward flow of the Subantarctic Mode Water (SAMW). Between 500 and 1200 m, the Antarctic Intermediate Water (AAIW) dominates the region (Table 3.1, Fig. 5.1) (Piola and Matano 2001). It is the major contributor to the upper layer return flow to the North Atlantic (de las Heras and Schlitzer 1999; You 2002). Several regions of enhanced mixing and modification of the AAIW have been identified within the South Atlantic, generally associated with small-scale mixing (Bianchi et al. 1993, 2002) and meso-scale eddies (Boebel et al. 1999); being the Brazil–Malvinas Confluence one of the most relevant regions where the exchange between relatively cold–fresh AAIW derived from Drake Passage and recirculated-AAIW occurs (Piola and Gordon 1989). North of the BMC, water depths between ~1700 and 3000 m are bathed by the southward flowing North Atlantic Deep Water (NADW), while to the south this water-mass encounters and interacts with the Circumpolar Deep Water (CDW), splitting it in two branches: the Upper CDW (between ~1000 and 2000 m depth) and the Lower CDW (between ~3000 and 3700 m depth) (Table 3.1, Fig. 5.3) (Reid et al. 1977; Saunders and King 1995; Piola and Matano 2001; Arhan et al. 2002, 2003; Henrich et al. 2003; Preu et al. 2013).

The most prominent feature of this region in determining the oceanographic structure of the western South Atlantic is the Antarctic Circumpolar Current (ACC). Currently, it transports 140 Sv (1 Sv = 10^6 m^3 seg^{-1}) (Siedler et al. 2001), mixing nutrient-rich and well-oxygenated waters from the South Atlantic with the poor-oxygenated waters from the Indian Ocean. Within this process, the CDW, AAIW and AABW have their origin. The shallow ones (CDW and AAIW) play a key role in the interhemispheric heat exchange (Lumpkin and Speer 2007) and control the nutrient-quantity supply to the Argentine margin through the Malvinas Current. The AAIW, Upper CDW and NADW boundaries can be distinguished by oxygen and nutrient content (Fig. 5.4). Whereas the AAIW is a high-oxygen and low-nutrient water-mass, the NADW is a nutrient-poor water-mass characterized by relatively homogeneous nitrate, phosphate and oxygen profiles (Reid et al. 1977). The CDW, by comparison, is an oxygen-poor and nutrient-rich water-mass with an upper branch characterized by higher phosphate, nitrate and silicate values between 900 m (tropical ocean) to 1500 m (subtropical ocean) (Stramma and England 1999). Below ~3700–4000 m depth, the northward flowing, corrosive Antarctic Bottom Water (AABW) fills the Argentine Basin (Arhan et al. 2002, 2003; Frenz et al. 2004), being the coldest, densest and deepest water-mass. It is characterized by very low temperature, salinity and nutrients, but high oxygen values (Reid et al. 1977;

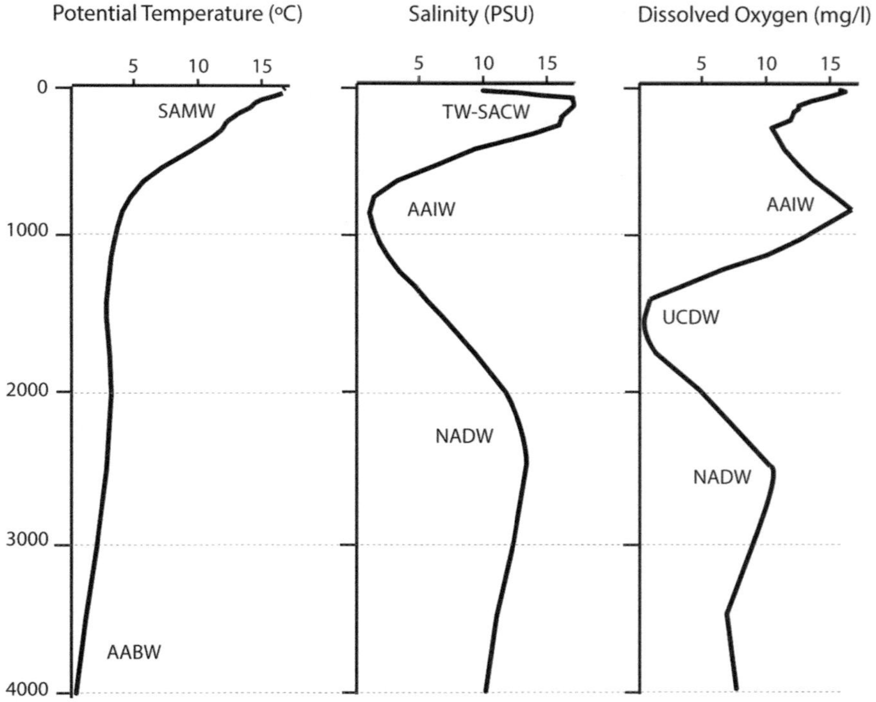

Fig. 5.4 Typical vertical profiles with CTD data from the Argentina Continental Margin (modified after García Chapori 2015). *Vertical scale* Meters water depth. *SAMW* Subantarctic Mode Water. *TW* Tropical Water. *SACW* South Atlantic Central Water. *AAIW* Antarctic Intermediate Water. *UCDW* Upper Circumpolar Deep Water. *NADW* North Atlantic Deep Water. *AABW* Antarctic Bottom Water

Alleman et al. 2001). The boundary between the Lower CDW and the AABW determines the depth of the calcite lysocline in the western South Atlantic (Reid et al. 1977; Frenz et al. 2004).

References

Alleman LY, Church TM, Véron AJ, Kim G, Hamelin B, Flegal AR (2001) Isotopic evidence of contaminant lead in the South Atlantic troposphere and surface waters. Deep Sea Res Part II 48 (13):2811–2827

Arhan M, Naveira Garabato AC, Heywood K, Stevens DP (2002) The Antarctic Circumpolar Current between the Falkland Islands and South Georgia. J Phys Oceanogr 32(6):1914–1931

Arhan M, Mercier H, Park YH (2003) On the deep water circulation of the eastern South Atlantic Ocean. Deep Sea Res Part I 50(7):889–916

Bianchi AA, Garzoli SL (1997) Variabilidad y movimientos del frente Brasil-Malvinas/Variability and motion of the Brazil-Malvinas front. Geoacta 22:74–90

Bianchi C, Gersonde R (2002) The Southern Ocean surface between Marine Isotope Stages 6 and 5d: shape and timing of climate changes. Palaeogeogr Palaeoclimatol Palaeoecol 187 (1):151–177

Bianchi AA, Giulivi CF, Piola AR (1993) Mixing in the Brazil-Malvinas confluence. Deep Sea Res Part I 40(7):1345–1358

Bender VB, Hanebuth TJJ, Chiessi CM (2013) Holocene shifts of the Subtropical Shelf Front off southeastern South America controlled by high and low latitude atmospheric forcings. Paleoceanography 28:1–10

Boebel O, Davis RE, Ollitrault M, Peterson RG, Richardson PL, Schmid C, Zenk W (1999) The intermediate depth circulation of the western South Atlantic. Geophys Res Lett 26:21

Boltovskoy E (1981) Masas de agua en el Atlántico Sudoccidental. In: Boltovskoy D (ed) Atlas de Zooplancton del Atlántico Sudoccidental (Mar del Plata), pp 227–236

Boltovskoy E, Boltovskoy D, Correa N, Brandini F (1996) Planktic foraminifera from the southwestern Atlantic (30–60 S): species-specific patterns in the upper 50 m. Mar Micropaleontol 28(1):53–72

Bryden HL, Imawaki S (2001) Ocean heat transport. In: Siedler G, Church J, Gould J (ed) Ocean Circulation and Climate, Elsevier New York: 455–474

Carvalho L, Jones C, Liebmann B (2004) The South Atlantic Convergence Zone: intensity, form, persistence, and relationships with intraseasonal to interannual activity and extreme rainfall. J Clim 17:88–108

Chiessi CM, Ulrich S, Mulitza S, Pätzold J, Wefer G (2007) Signature of the Brazil-Malvinas Confluence (Argentine Basin) in the isotopic composition of planktonic foraminifera from surface sediments. Mar Micropaleontol 64(1):52–66

Chiessi CM, Mulitza S, Groeneveld J, Silva JB, Campos MC, Gurgel MH (2014) Variability of the Brazil Current during the Late Holocene. Palaeogeogr Palaeoclimatol Palaeoecol 415:28–36

Clapperton C (1993) Quaternary geology and geomorphology of South America. Elsevier Sciences Publication, New York 779 pp

Combes V, Matano RP (2014) Trends in the Brazil/Malvinas Confluence region. Geophys Res Lett 41(24):8971–8977

de las Heras M, Schlitzer R (1999) On the importance of intermediate water flows for the global ocean overturning. J Geophys Res Oceans 104:15515–15536

Dima I, Wallace JM (2003) On the seasonality of the Hadley Cell. J Atmos Sci 60:1522–1527

Fetter AF, Matano RP (2008) On the origins of the variability of the Malvinas Current in a global, eddy-permitting numerical simulation. J Geophys Res Oceans (1978–2012) 113(C11)

Figueroa SN, Satyamurty P, da Silva Leite, Dias P (1995) Simulations of the summer Circulation over the South American Region with the ETA coordinate model. Am Meteorol Soc 52:1573–1584

Flantua SGA, Hooghiemstra H, Vuille M, Behling H, Carson JF, Gosling WD, Hoyos I, Ledru MP, Montoya E, Mayle F, Maldonado A, Rull V, Tonello MS, Whitney BS, Gonzalez-Arango C (2015) Climate variability and human impact on the environment in South America during the last 2000 years: synthesis and perspectives. Clim Past Discuss 11:3475–3565. doi:10.5194/cpd-11-3475-2015

Frenz M, Höppner R, Stuut J-BW, Wagner T, Henrich R (2004) Surface sediment bulk geochemistry and grain-size composition related to the oceanic circulation along the South American Continental Margin in Southwest Atlantic. In: Wefer G, Mulitza S, Ratmeyer V (eds) The South Atlantic in the Late Quaternary. Springer, pp 347–373

Fu R, Zhu B, Dickinson RE (1998) How do atmosphere and land surface influence seasonal changes of convection in the tropical Amazon? J Clim 12:1306–1321

Gappa JL (2000) Species richness of marine Bryozoa in the continental shelf and slope off Argentina (south-west Atlantic). Divers Distrib 6(1):15–27

García Chapori N (2015) Reconstrucción paleoceanográfica del talud bonaerense a partir de testigos cuaternarios. Ph.D. thesis, University of Buenos Aires, 203 pp

García Chapori N, Chiessi CM, Bickert T, Laprida C (2015) Sea-surface temperature reconstruction of the Quaternary western South Atlantic: new planktonic foraminiferal correlation function. Palaeogeogr Palaeoclimatol Palaeoecol 425:67–75
Garreaud RD, Vuille M, Compagnucci R, Marengo J (2009) Present-day South American climate. Palaeogeogr Palaeocl 281:180–195. doi:10.1016/j.palaeo.2007.10.032
Garrison T (2008) Essentials of oceanography. Brooks/Cole Belmont, CA, USA. ISBN-13: 978-0-8400-6155-3, ISBN-10: 0-8400-6155-2
Garzoli SL, Giulivi C (1994) What forces the variability of the southwestern Atlantic boundary currents? Deep Sea Res Part I 41(10):1527–1550
Garzoli SL, Matano R (2011) The South Atlantic and the Atlantic meridional overturning circulation. Deep Sea Res Part II 58:1837–1847
Goni GJ, Wainer I (2001) Investigation of the Brazil Current front variability from altimeter data. J Geophys Res Oceans (1978–2012) 106(C12):31117–31128
Goni GJ, Bringas F, DiNezio PN (2011) Observed low frequency variability of the Brazil Current front. J Geophys Res Oceans (1978–2012) 116(C10)
Gordon AL (1981) South Atlantic thermocline ventilation. Deep Sea Res Part A Oceanogr Res Pap 28(11):1239–1264
Gordon AL (1989) Brazil-Malvinas Confluence-1984. Deep Sea Res 36:359–384
Henrich R, Baumann K-H, Gerhardt M, Gröger M, Volvers A (2003) Carbonate preservation in deep and intermediate water masses in the South Atlantic: evaluation and geological record (A review). In: Wefer G, Mulitza S, Ratmeyer V (eds) The South Atlantic in the Late Quaternary: reconstruction of material budgets and currents systems. Springer, Berlin, pp 645–670
Horel J, Hahmann A, Geisler J (1989) An investigation of the annual cycle of the convective activity over the tropical Americas. J Clim 2:1388–1403
Jeffreys H (1925) On the formation of water waves by wind. Proc R Soc Lond 107A:189–206
Kodama Y (1992) Large-scale common features of subtropical precipitation zones (the Baiu Frontal Zone, the SPCZ, and the SACZ) Part I: characteristics of subtropical frontal zones. J Meteorol Soc Jpn 7:813–836
Kuhlbrodt T, Griesel A, Montoya M, Levermann A, Hofmann M, Rahmstorf S (2007) On the driving processes of the Atlantic meridional overturning circulation. Rev Geophys 45 (RG2001)
Laprida C, García Chapori N, Chiessi CM, Violante RA, Watanabe S, Totah V (2011) Middle Pleistocene sea surface temperature in the Brazil-Malvinas Confluence Zone: Paleoceanographic implications based on planktonic foraminifera. Micropaleontology 57 (2):183–194
Legeckis R, Gordon AL (1982) Satellite observations of the Brazil and Falkland currents—1975, 1976 and 1978. Deep Sea Res Part A Oceanogr Res Pap 29(3):375–401
Liebmann B, Kiladis G, Marengo J, Ambrizzi T, Glick JD (1999) Submonthly convective variability over South America and the South Atlantic Convergence Zone. J Clim 11:2898–2909
Lumpkin R, Speer K (2007) Global ocean meridional overturning. J Phys Oceanogr 37(10):2550–2562
Lutjeharms JRE, Meeuwis JM (1987) The extent and variability of South-East Atlantic upwelling. S Afr J Mar Sci 5(1):51–62
Marengo J, Liebmann B, Kousky VE, Filizola N, Wainer IC (2001) Onset and end of the rainy season in the Brazilian Amazon Basin. J Clim 14:833–852
Matano RP (1993) On the separation of the Brazil Current from the coast. J Phys Oceanogr 23 (1):79–90
Matano RP, Palma ED (2008) On the upwelling of downwelling currents. J Phys Oceanogr 38 (11):2482–2500
Matano RP, Schlax MG, Chelton DB (1993) Seasonal variability in the southwestern Atlantic. J Geophys Res Oceans (1978–2012) 98(C10):18027–18035
Matano RP, Palma ED, Piola AR (2010) The influence of the Brazil and Malvinas Currents on the southwestern Atlantic shelf circulation. Ocean Sci 6(4):983–995

Möller OO, Piola AR, Freitas AC, Campos EJ (2008) The effects of river discharge and seasonal winds on the shelf off southeastern South America. Cont Shelf Res 28(13):1607–1624
Munk WH, Wunsch C (1998) Abyssal recipes II: energetics of tidal and wind mixing. Deep Sea Res 45:1977–2010
Nakamura H, Shimpo A (2004) Seasonal variations in the southern hemisphere storm tracks and jet streams as revealed in a reanalysis dataset. J Clim 17:1828–1844
Olson DB, Podestá GP, Evans RH, Brown OB (1988) Temporal variations in the separation of Brazil and Malvinas Currents. Deep Sea Res Part A Oceanogr Res Pap 35(12):1971–1990
Palma ED, Matano RP, Piola AR (2008) A numerical study of the Southwestern Atlantic Shelf circulation: Stratified ocean response to local and offshore forcing. J Geophys Res Oceans (1978–2012) 113(C11)
Peterson RG, Stramma L (1991) Upper-level circulation in the South Atlantic Ocean. Prog Oceanogr 26:1–73
Piola AR, Gordon AL (1989) Intermediate waters in the southwest South Atlantic. Deep Sea Res Part A Oceanogr Res Pap 36(1):1–16
Piola AR, Matano RP (2001) Brazil and Falklands (Malvinas) currents. Ocean currents: a derivative of the encyclopedia of ocean sciences, pp 35–43
Piola AR, Campos EJD, Möller OO, Charo M, Martínez C (2000) Subtropical shelf off eastern South America. J Geophys Res 105:6565–6578
Piola AR, Möller OO, Guerrero RA, Campos EJ (2008) Variability of the subtropical shelf front off eastern South America: winter 2003 and summer 2004. Cont Shelf Res 28(13):1639–1648
Piola AR, Avellaneda NM, Guerrero RA, Jardón FP, Palma ED, Romero SI (2010) Malvinas-slope water intrusions on the northern Patagonia continental shelf. Ocean Sci 6(1):345
Preu B, Hernández-Molina FJ, Violante RA, Piola AR, Paterlini CM, Schwenk T, Voigt I, Krastel S, Spieß V (2013) Morphosedimentary and hydrographic features of the northern Argentine margin: the interplay between erosive, depositional and gravitational processes and its conceptual implications. Deep Sea Res 75:157–174
Reboita MS, Gan MA, da Rocha RP, Ambrizzi T (2010) Regimes de precipitação na América do sul: uma revisão bibliográfica. Rev Bras Meteorologia 25:185–204
Reid JL, Nowlin WD Jr, Patzert WC (1977) On the characteristics and circulation of the southwestern Atlantic Ocean. J Phys Oceanogr 7(1):62–91
Robertson AW, Mechoso CR (2000) Interannual and interdecadal variability of the South Atlantic Convergence Zone. Mon Weather Rev 128:2947–2957
Sandström JW (1916) Meteorologische studien à schedischen hochgebirge. Göteborgs K. Vetensk. Vitterhets-Samh Handl 10:2–42
Saunders PM, King BA (1995) Oceanic fluxes on the WOCE A11 section. J Phys Oceanogr 25 (9):1942–1958
Schloesser F, Furue R, McCreary JP, Timmermann A (2012) Dynamics of the Atlantic meridional overturning circulation. Part 1: Buoyancy-forced response. Prog Oceanogr 101(1):33–62
Shannon LV, Agenbag JJ, Walker ND, Lutjeharms JRE (1990) A major perturbation in the Agulhas retroflection area in 1986. Deep Sea Res Part A Oceanogr Res Pap 37(3):493–512
Siedler G, Gould J, Church JA (2001) Ocean circulation and climate: observing and modelling the global ocean, vol 103. Academic Press
Solman SA (2013) Regional climate modeling over South America: a review. Advances in Meteorology 2013, Article ID 504357. doi:10.1155/2013/504357
Stramma L, England M (1999) On the water masses and mean circulation of the South Atlantic Ocean. J Geophys Res Oceans (1978–2012) 104(C9):20863–20883
Talley LD (2003) Shallow, intermediate and deep overturning components of the global heat budget. J Phys Ocean 33:530–560
Toggweiler JR, Samuels BL (1993) Is the magnitude of the deep outflow from the Atlantic Ocean actually governed by Southern Hemisphere winds? In: Heimann M (ed) The global carbon cycle. Springer, Berlin, pp 303–331
Toggweiler JR, Samuels BL (1995) Effect of Drake passage on the global thermohaline circulation. Deep Sea Res Part I 42:477–500

Toggweiler JR, Samuels BB (1998) On the Ocean large-scale circulation near the limit of no vertical mixing. J Phys Oceanogr 28:1832–1852
Toledo FA, Costa KB, Pivel MA (2007) Salinity changes in the western tropical South Atlantic during the last 30 kyr. Global Planet Change 57(3):383–395
Toledo FA, Costa KB, Pivel MA, Campos EJ (2008) Tracing past circulation changes in the western South Atlantic based on planktonic foraminifera. Rev Bras Paleontol 11(3):169–178
Vera C, Higgins W, Amador J, Ambrizzi T, Garreaud R, Gochis D, Gutzler D, Lettenmaier D, Marengo J, Mechoso CR, Nogues-Paegle J, Silva Diaz PL, Zhang C (2006) Towards a unified view of the American Monsoon System. J Clim 19:4977–5000
Voigt I, Chiessi CM, Prange M, Mulitza S, Groeneveld J, Varma V, Henrich R (2015) Holocene shifts of the southern westerlies across the South Atlantic. Paleoceanography 30(2):39–51
Wainer I, Gent P, Goni G (2000) Annual cycle of the Brazil-Malvinas Confluence region in the National Center for Atmospheric Research climate system model. J Geophys Res Oceans (1978–2012) 105(C11):26167–26177
Wilson HR, Rees NW (2000) Classification of mesoscale features in the Brazil-Falkland Current confluence zone. Prog Oceanogr 45(3):415–426
You Y (2002) Quantitative estimate of Antarctic Intermediate Water contributions from the Drake Passage and the southwest Indian Ocean to the South Atlantic. J Geophys Res Oceans (1978–2012) 107(C4):6-1
Zhou J, Lau KM (1998) Does a monsoon climate exist over South America? J Clim 11:1020–1040

Chapter 6
Principles of Paleoceanographic Reconstruction

Abstract This chapter introduces the principles of paleoceanographic reconstructions and proxy data, focusing on the concepts of climatic, biological, geological, geochemical as well as other large-scale proxies, tracers and records useful for such reconstructions in the western South Atlantic. Different proxies particularly useful for the Argentine margin, including physical and chemical properties of sediments, microfossils and geochemical and isotopic properties, are described.

Keywords Paleoceanographic reconstruction · Proxies · Tracers · Records · Biological proxies · Geological proxies · Geochemical proxies

Oceanographic processes are closely tied to climate (e.g., Wefer and Berger 1996; Berger et al. 2010) in the framework of the ocean-atmosphere coupling that regulates the exogene processes at sea and the sedimentary, geochemical and biological processes occurring there. Oceans play a major role in the climate system, as they are responsible for a permanent exchange of heat, water, carbon dioxide and other elements with the atmosphere. This exchange incides in the climate variability. Therefore, some changes appropriately measured in the marine sediments and faunas, as indicators of oceanographic changes, can directly or indirectly be related to climate changes. In this way, we have first at all to refer to paleoclimatology as the ultimate stage of paleoreconstructions in the sea realm. Paleoclimatology is the study of climate preceding the period of instrumental measurements. Meteorological observations are limited in space and time, and as such, instrumental records span only a small fraction ($<10^{-7}$) of the Earth's history. Due to the nature of the Earth Climate System, several biological, geochemical, biogeochemical and geological processes are dependent on climate and incorporate into their structure a measure of this climatic dependency. If these processes are somehow preserved in the geological record, an extended perspective on climatic variability can be obtained by the study of these natural phenomena climate-dependent.

R.A. Violante et al., *The Argentina Continental Margin*,
SpringerBriefs in Earth System Sciences, DOI 10.1007/978-3-319-04196-4_6

6.1 Paleoceanography and Proxy Data

Past oceanographic-climatic-environmental conditions can be interpreted from natural elements (or phenomena) that leave certain useful signals in the geological record. These signals can be referred to as "proxy data" which stand as surrogates for particular climate variables. Based essentially on biological and geological proxy information, the climate that prevailed before humans made possible to record climatic history can be in this way reconstructed. For example, oceanic temperatures and ice volume (climatic parameter) impact on the distribution of environmentally stable isotopes such as oxygen-18 (^{18}O), hydrogen-2 (deuterium- ^{2}H) and carbon-13 (^{13}C) in the marine realm. Thus, stable isotopes of C and O of the biogenic carbonate of fossil planktonic foraminifera shells (proxy data) provide an indirect record of ocean-water temperatures, salinity, changes in the global ice-volume (ice-sheet growth and decay) or regional changes in the hydrological cycle.

Therefore, the proxy materials, if conveniently preserved, constitute a more or less permanent record representing past climatic conditions. However, that record can incorporate other signals related to non-climatic influences. Studies of past climates must begin with an understanding of the proxy data available in the geological record and the methods and assumptions used in their analysis, including the isolation of the climatic and non-climatic signals. It may then be possible to synthesize different lines of evidence into a comprehensive picture of former climatic fluctuations, and to test hypotheses about the causes of past climate changes.

For extracting the paleoclimatic signal from proxy data, the record must first be calibrated. Calibration involves modern climatic records and modern proxy data to understand how, and to what extent, proxy data are climate-dependent. It is assumed that the modern relationships observed in the climatic system have operated, unchanged, throughout the period of study. However, even when paleoclimatic research is built on studies of climate dependency in natural phenomena operating today, not all environmental/climatic conditions in the past are represented at the present time. For example, circulation in the Cretaceous ocean may have been very different from that of today. The present ocean structure depends on constant wind systems. During most of the Cretaceous the polar regions were ice-free. Without polar ice there were seasonal reversals in the high-latitude atmospheric pressure systems. Without constant mid-latitude westerly winds, there would be no subtropical and polar fronts in the ocean, no well-developed ocean pycnocline, and no tropical-subtropical gyres dominating ocean circulation (Hay 2008). One must therefore be aware of the possibility that erroneous paleoclimatic reconstructions may result from the use of modern climate-proxy data relationships when past conditions have no analog in the modern world.

The proxy data actually comprise a wide range of elements varying from mega (large-scale) to micro (low-scale) components of sediments, such as those indicated in Table 6.1. Those elements can be originated in two kind of processes (Goudie 2004): episodic/discontinuous (i.e., catastrophic) events that occur from quasi instantaneous natural phenomena that result from the integration of climatic (and/or

Table 6.1 Proxies useful for Late Quaternary paleoreconstructions

Systems	Source	Proxy		Environmental parameter
Biological	Foraminiferida	Planktonic	Species composition, oxygen isotopic ratio	Sea-surface temperature, water-masses, productivity, upwelling, thermocline depth, sea-surface salinity, lysocline depth, seasonality
		Benthic	Species composition, oxygen and carbon isotopic ratio	Temperature, water-masses, productivity, paleobathymetry, sea level, stratigraphy, bottom-water oxygenation, lysocline depth
	Corals	Deep-sea stony corals	Species composition, isotopic ratios	Water-masses, ice-volume, oxygenation, temperature
		Coral reefs	Species composition, isotopic ratios	Nutrient, sea level, salinity, ice-volume, temperature
	Actinopoda	Radiolaria	Species composition	Sea-surface temperature, salinity, water-masses
	Chrysophyta	Diatoms	Species composition, C, N and Si isotope geochemistry	Sea-surface temperature, productivity, sea-ice cover
	Dinoflagellata	Organic-walled cysts	Species composition	Sea-surface temperature, productivity, sea-ice cover, salinity, sea level
	Crustacea	Ostracods	Species composition, isotopic ratios	Water-masses, salinity paleobathymetry, oxygenation
	Haptophyta	Coccolithophorids	Species composition, biomarkers	Sea-surface temperature, productivity
	Mollusk		Species composition, isotopic ratios	Temperature, salinity, sea level

(continued)

Table 6.1 (continued)

Systems	Source	Proxy		Environmental parameter
Geochemical	Stable isotopes	Oxygen	$^{18}O/^{16}O$	Temperature, ice-volume
		Carbon	$^{13}C/^{12}C$	Productivity
		Strontium	$^{87}Sr/^{86}Sr$	Salinity, continental runoff, vulcanism
		Boron	$^{10}B/^{11}B$	pH
	Radiogenic isotopes	Os, Nd, Sr, Pb, Th, Pa	^{87}Sr, ^{187}Os, ^{143}Nd, ^{206}Pb, ^{207}Pb, ^{208}Pb, $^{231}Pa/^{230}Th$	Water-masses, circulation patterns, flow rates
	Elemental composition	Element ratios	Mg/Ca, U/Ca, Sr/Ca and Cd/Ca in carbonate shells Ti/Sr in terrigenous/biogenic materials	Temperature, productivity, source, terrigenous versus hemipelagic rate
		Trace elements	Siliciclastic signature: Al, Ti, Fe	Changes in terrigenous input, terrigenous versus hemipelagic sedimentation, sea level, glaciation–deglaciation
			Biogenic signature: Ca, Sr	Changes in terrigenous input, terrigenous versus hemipelagic sedimentation sea level, glaciation–deglaciation
			B, Nd, Ga, Rb, Sr	Salinity, sediment source
			Mn, Cr, Re, Cd, Co	Redox conditions
			Ba, P, Al	Productivity
	Biomarkers	C_{37} Alkenones	Alkenone unsaturation index (U^{k}_{37})	Sea-surface temperature, stratigraphy
		N-alkyl-lipids	Tex $_{86}$ index	Sea-surface temperature

(continued)

Table 6.1 (continued)

Systems	Source	Proxy		Environmental parameter
Physical	Mineralogy	Siliciclastics	Siliciclastic (terrigenous) minerals	Source, changes in terrigenous input, sea level, glaciation–deglaciation, sediment reworking
		Carbonates	Limestone, dolomite, chalk	Temperature
		Clay minerals	Illite, chlorite, kaolinite, smectite	Deep circulation, current velocities, water-mass origin, weathering conditions, provenance and source area
	Sedimentology	Sediment texture	Grain-size, sorting	Source, sea-ice cover, currents speed and direction, depositional conditions, depositional-erosional processes
		Sediment structures	Stratification, lamination. Sequence stratigraphy, progradation, retrogradation	Current velocity, storms, flow regime, gradational processes, terraces and contouritic bodies, turbidites and mass-transport deposits
	Geomorphology	Depositional and erosional features	Contouritic terraces and channels, moats, turbidites, submarine canyons	Circulation, alongslope currents and across-slope processes
		Surface morphosedimentary evidence	Ancient shorelines, contouritic terraces, slope, submarine canyons, gravitational features, deep-sea fans	Sea level, deep-sea along- and acrossslope processes, depositional-erosional features, contourites, turbidites, mass-transport deposits
	Paleomagnetic properties	Natural remanent magnetization	NRM	Mineralogy, grains size of magnetic minerals, intensity and direction of the geomagnetic field
		Isothermal remanent magnetization	IRM	Concentration of ferrimagnetic mineral
		Anhysteretic remanent magnetization	ARM	Concentration of ferrimagnetic mineral
		Magnetic susceptibility	k	Amount of ferromagnetic minerals, mineralogy, dissolved oxygen, flow speed

others) conditions prior to the event (e.g., flooding, slides, glacial advances, etc.); and continuous/incremental processes (e.g., sedimentation, biogenic accumulations) which preserve quasi continuous records of environmental/climatic conditions spanning a relatively large time-frame.

As such a reconstruction depends on the environmental conditions, in the case of the oceans different methodological approaches will be needed according to the studied environment, the predominant processes acting on it, and the kind of proxy data available (biological/geological, large/medium/low-scale). For example, the shelf and adjacent coasts were deeply affected by sea-level fluctuations with alternately exposures to subaerial processes and flooding (in response to the shift from glacial to interglacial conditions), as well as to higher or lower energy dynamic processes; therefore, approaches such as sequence stratigraphy, sedimentology and geochemical records indicating changes from marine to continental conditions are among the most important proxies to consider. On the other hand, the deepest regions (slope, rise and deep-ocean environments) were never exposed to subaerial conditions and were not directly affected by sea-level fluctuations; therefore they do not show alternation of marine and continental features and they are more sensitive to changes in water temperature, salinity and density, as well as fluctuations in the position of interfaces between adjacent water-masses, changes in the predominance of alongslope versus downslope processes and variations in the direction of circulation of submarine currents; in the case of these deeper environments, biological and geochemical proxies are, in general, more relevant.

Although "proxy" is a general term usually associated to any element—or signal—recording an environmental change, it is worth mentioning—following Hillaire-Marcel and de Vernal (2007)—that the different elements preserved in the sedimentary sequences that constitute the sea floor, either its megastructures or the microscopic components (lithological, chemical, and biogenic) can be considered as proxies, tracers or records according to the scale and the degree of information provided. Although the definition of each of them is not strict and their meanings can overlap, the following differences can be stated:

Proxy: A measurable property of an environmental/geological record which, through mathematical or statistical treatment, can be related—with a stated uncertainty—to one or a combination of physical, chemical or biological environmental factors during its formation. Therefore, the proxy provides a quantitative estimate of the environment.

Tracer: Environmental/geological properties or objects (entities) that cannot be adequately measured and therefore they provide quantitative or qualitative information about conditions or processes.

Record: Physical entity (or object) that preserves properties linked to the past environmental/geological conditions that have not been modified (at least at a measurable level) by diagenetic changes. Some authors consider "records" as large-scale proxies, and particularly McCave (2007) stresses the importance of the mega-sedimentary evidences (sedimentary structure and architecture) as a valuable tool for deciphering sedimentary evolution, direction and velocity of sediment transport, and other large-scale processes without which the micro-attributes (biological content, geochemistry, isotopic signatures, etc.) cannot be adequately interpreted.

6.2 Climate Proxies

Climate proxies can be classified into physical, biological or geochemical according to the available element to be studied. In paleoceanography they include physical properties of the sediments (grain-size, mineralogy, density and magnetic properties), geochemical (including isotopic) properties of detrital and biogenic materials, and fossils preserved as calcium carbonate, silica, refractory organic matter and organic biomarkers. Some of them should provide quantitative values of the reconstructed parameter, whereas others (commonly referred to as tracers) are related to environmental conditions and may provide quantitative or qualitative insight into such conditions. The list included in Table 6.1 is undoubtedly incomplete and not necessarily may conform to every user. New methodologies are emerging all the time and old methodologies are being rediscovered and reinterpreted continuously. The emphasis in this classification is, on first principle, applied to paleoclimatology rather than on components where they are embedded. For example, fossil foraminiferal assemblages from marine sediments carry several different types of climatic signals: specific composition and proportion of species (assemblage structure), isotopic and/or Mg/Ca ratios of calcareous shells can be climatically interpreted under different disciplinary principles and assumptions. Each line of evidence differs in its ability to resolve events accurately and in the specific time-frame of application. To reconstruct past climate variability and oceanic conditions (for instance sea-ice cover, surface or bottom water temperature, nutrient concentrations, biological productivity), marine geologists have to rely on the analysis of particular marine proxies. The main carriers of paleoclimate and paleoceanographic information are the inorganic or organic preserved remains of microfossils: foraminifers (calcareous zooplankton and zoobenthos), coccolithophores (calcareous algae, phytoplankton), diatoms (Bacillariophyceae, siliceous algae) as well as radiolarians and silicoflagellates (siliceous zooplankton). Corals (Cnidaria) and other macro-invertebrates such as bivalves (Pellecipoda, Mollusca), are commonly used as climate proxies indeed. Paleobiological aspects of fossil assemblages such as population dynamics, relative abundance as well as species composition of assemblages are frequently used to infer past climate and environmental conditions whereas organic or inorganic geochemical aspects can reflect water chemistry at the time of biomineralization or growth. For instance, the occurrence of warm-water taxa *Globorotalia menardii* at higher latitudes of present-day distribution is used as evidence for warmer sea-surface temperatures. The unsaturation ratio of unsaturated C_{37} alkenones, an organic molecular fossil derived from coccolithophore, is an excellent proxy for quantitative sea-surface temperature.

Chemical and physical properties of marine sediments may also be useful for paleoclimate reconstructions. Variations in composition, grain-size and flux of marine sediments hold clues to important paleoclimate variables such as wind speed and direction, aridity, glacial activity and ocean currents. The distribution (and dissolution) of carbonates in the deep-sea is ultimately controlled by the saturation state of the CO_2 in the mean ocean water, which itself is a function of the climate. Thus, mineralogy, petrography, grain-size distribution and geochemical approaches (including isotopes) of marine sediments allow the reconstruction of sediment provenance and sedimentary processes, weathering history, flux, sorting and changes in atmospheric CO_2 and the global carbon cycle.

Different proxies have different levels of sensitivity with respect to climate: some systems vary in-phase with climatic variations whereas others lag behind by as much as several centuries or even millennia. The record of planktonic foraminifera, for example, may point to short-term changes of climate because foraminiferal populations are often highly sensitive to temperature fluctuations. By contrast, ice-volume takes up to hundred years to adjust to an abrupt change in climate. This is not a question of dating accuracy but a fundamental attribute of the proxy system analyzed. Additionally, not all proxies provide a continuous record of climate. Certain phenomena provide discontinuous or episodic information, whereas other records provide continuous paleoclimatic signals in a more or less extended time series. Dropstones, for example, may leave evidence of sea-ice extent but the processes of their deposition represent a discrete event in time, while marine sediments may provide continuous records for over several millions years, and hence they are commonly used as a paleoclimatic frame of reference for long-term climatic fluctuations.

In all climatic or environmental reconstructions, precise dating is crucial: without accurate dating it is impossible to determine the precise age of an event or the temporal relation between spatially distant events. Errors associated with multiple dating methods must be carefully considered. This is a fundamental requirement if we are to understand the nature and structure of global changes of the past, particularly when considering high frequency, short-term changes in climate.

Multi-proxy studies, where several proxies are studied on the same stratigraphic sequence, are particularly important as they enable the testing of alternative and complementary hypotheses to explain the observed changes. The use of simulation models is another important approach to testing alternative hypotheses about underlying factors for the observed changes.

6.3 Biological Proxies: Fossils as Paleoceanographic Proxies

The paleoceanographic interpretation of fossils is essentially based on analogy with their living counterparts. Many organisms appear to have restricted ecologic, biogeographic and/or bathymetric distributions rendering them of utility in environmental reconstructions. Ecological requirements and tolerances of foraminifers, corals, radiolarians, diatoms, dinoflagellates, ostracods, coccoliths and bivalves, and their rapid response to changing environmental conditions, render them useful in paleoceanography and paleoclimatology.

6.3.1 *Foraminiferida*

Foraminifers (Rhizaria, Actinipoda) are probably the most important source of paleoceanographic information. Foraminifera are marine heterotrophic Protista. Attributes of foraminifer's shells and fossil assemblages make them ideally suited to record paleoenvironmental changes. Foraminifer's shells are composed of calcium carbonate, incorporating oxygen, carbon and boron isotopes, replacing the calcium and incorporating trace metals into the calcite lattice. As a result, carbonate shells are the substrate for extraction of several geochemical signals. Additionally, foraminifera are abundant, ubiquitous and ecologically sensitive, and fossil foraminifera assemblages are generally well preserved and highly informative tools concerning physicochemical past conditions.

The distribution and abundance of planktonic foraminifera species is strongly linked to surface-water properties, and as such, extant species can be grouped into five main assemblages that define the Tropical, Subtropical, Temperate, Subpolar and Polar provinces (Boltovskoy 1981). In turn, benthic foraminiferal tests are carriers of geochemical proxies, and benthic foraminiferal assemblages have been used to track variations in the extension of deep-water masses and deep-sea circulation patterns, and more recently, bottom water oxygenation and various aspects of the organic flux to the ocean floor.

6.3.2 *Coccolithophorids*

Coccolithophores (Chromista, Haptophyta) are unicellular, autotrophic small organisms belonging to marine phytoplankton. During certain phases of their life cycle, they are able to produce minute calcium carbonate plates generally smaller than 5 μm called coccoliths. The living cells, restricted to the upper part of the water column, are characterized by having a golden brown chloroplast. As a consequence of their small size, they are included in calcareous nannoplancton. Light intensity is the main limiting factor for coccolithophores, although temperature, nutrient contents or salinity also control their distribution. Large blooms of these algae are observed covering extensive regions of the oceans. The relationship between the coccolithophores assemblages and environmental parameters, particularly sea-surface temperature, the stability and thickness of the mixed layer and nutrient content (and position of the nutricline), makes it possible to use coccoliths as paleoceanographic proxies. They provide information about sea-surface temperature, water-masses, nutrient availability and productivity, as well as atmospheric processes related with mixed layer stability.

6.3.3 *Dinoflagellates*

Dinoflagellates (Dinophyta, Dinoflagellata) are single-celled organisms generally between 5 and 2000 μm in size with both animal and plant characteristics, since heterotrophic and autotrophic modes of nutrition occur. Dinoflagellates have two flagella for propulsion and autotrophic forms constitute an important part of the oceanic phytoplankton living in the photic zone where trace elements availability limits their productivity. On a global scale modern dinoflagellates occupy broad latitudinal zones encompassing low-, middle—and high-latitudes. Ocean currents influence the distribution of dinoflagellate cysts and the marine microplankton as a whole. The majority exhibits alternation of generations in the life cycle, but it is the resistant resting cyst which leaves a fossil record. The cyst wall is built of organic material resistant to bacterial decay. Dinoflagellate cysts have proved to be valuable tools in paleoceanography. The species composition of the assemblages appears informative at different levels: sea-level changes, sea-surface conditions such as sea-ice cover, temperature, productivity as well as salinity and hydrological fronts.

6.3.4 *Diatoms*

Diatoms (Protista, Bacillariophyta) are unicellular organisms in which the cell is encapsulated in a frustule composed of two highly ornamented amorphous silica valves. Diatom size varies from 2 μm to 1–2 mm. They are found in almost every aquatic environment, including fresh and marine waters. Diatoms are photosynthetic organisms possessing yellow–brown chloroplasts with chlorophyll and a large set of carotenoids pigments which enable diatoms to live at low light levels, for example under sea-ice. In the marine environment diatoms are generally planktonic, nonmotile and restricted to the photic zone. Many factors interact to determine the distribution of planktonic diatoms, being the most important the sea-surface temperatures, nutrient levels, light level and stability of the mixed layer. Salinity may also exert a major role on diatom distribution where strong gradients in salinity exists, such as coastal zones and regions influenced by sea-ice. Where sediments are mainly barren of $CaCO_3$, diatoms are the prime paleoceanographic tool. Fossil diatom assemblages and key species can be used to track past environmental conditions based on ecological preferences, especially productivity, sea-surface temperature and sea-ice history. Transfer functions produce quantitative estimates of surface physicochemical parameters, such as sea-surface temperature (SST) and sea-ice cover.

Four isotope ratios are measured in the diatom organic-intrinsic matter (C and N) and in the diatom frustule (O and Si). Up until now, carbon, nitrogen and silicon isotopes are used to document past changes in productivity and nutrient cycling as well as in related water-mass fluxes. In turn, $\delta^{18}O$ of diatoms frustule is dependent upon the SST and the isotopic composition of the sea water, and equations linking diatom $\delta^{18}O$ and SST have been developed.

6.3.5 *Radiolarians*

The term radiolarian (Radiozoa) includes several closely related groups of unicellular organism between 50 and 400 μm in diameter characterized by the presence of a perforate organic membrane called central capsule, which separates the cytoplasm into an outer ectoplasm and an inner endoplasm. A mineralized skeleton of opaline silica is usually present within the cell. Skeleton comprises radial and/or tangential elements. The radial elements consist of hollow or solid spicules, spines and bars. The tangential elements, where present, generally form a porous lattice shell of very variable morphology. As marine zooplankton, radiolarians occupy a wide range of trophic types. They prefer open oceanic conditions, especially the nutrient-rich waters just seaward of the continental slope. The temperature, silica and other nutrient concentrations influence the latitudinal abundance of radiolarian, and as such, geographical distribution is directly related to ocean circulation and water-masses distribution. Most species are restricted to specific habitats. Radiolaria have value as depth, climate and temperature proxies. They serve as water-masses identification, paleotemperature and paleoproductivity estimations particularly where the calcareous microfossils have suffered dissolution.

6.3.6 *Corals*

Corals include a wide range of stony (Anthozoa, Hexacorallia) and horny (Anthozoa, Octocorallia) cnidarians. The understanding of the temporal and environmental significance of growth increments of their skeletons allows their use in paleoceanographic reconstructions.

Stony corals are polyp animals found exclusively in marine environments. They include solitary and colonial forms, many of which secrete a calcareous exoskeleton that provides support and protection to polypids. Corals can be reef-builders and non-reef-builders. The former inhabit shallow waters on tropical oceans. The geochemistry ($\delta^{18}O$) of their skeletons provides a record of temperature and salinity, and distribution of species allows reconstruction of sea-level elevations. Additionally, growth patterns reflect variations in environmental parameters such as water quality, light conditions and nutrient availability. Non-reef-builder corals are not restricted to shallow waters and inhabit all regions and depths of the ocean.

Geochemistry of skeletons encodes environmental variability. The ð^{18}O of coralline aragonite is primarily influenced by sea-surface temperature, and isotopic signature of seawater and elemental ratios (i.e., Mg/Ca, Sr/Ca) has been used to reconstruct environmental parameters. Distribution of some deep-sea species seems to be related with specific bottom water-masses.

Horny corals have received comparatively less attention, but several genera form durable skeletons. Skeletons exhibit a carbonate phase (usually high magnesium calcite) and a horny, protein-rich phase called gorgolin. Isotopic and amino acid chemistry of this horny phase is utilized in paleoceanography. The ð^{15}N is used for water-masses movements and trophic dynamics (i.e., complexity of food webs, uptake of phytoplankton). The ð^{13}C is used to analyze variability in the ð^{13}C of oceanic dissolved organic carbon reservoir (DIC).

6.3.7 Ostracods

Ostracods (Crustacea, Athropoda) are the most useful group of crustaceans in paleoceanographic studies. The most characteristic feature of their bodies is a bivalved, well-calcified shell between 0.3 and 2 mm length which consists of 80–90% calcite providing a good target for geochemical analyses. They live in brackish, saline and hypersaline transitional continental-marine environments, and at sea they are found from the shoreline down to hyperabyssal depths.

Ostracods have potential as paleoceanographic proxies because many species are ecologically sensitive and limited to particular oceanic conditions and benthic habits. Salinity and the nature of the substrate are the most fundamental factors determining the distribution of ostracods in the inner shelf, but deep-sea ostracods have complex (and still not fully understood) ecological requirements. Nutrient supply and depth-related parameters (i.e., temperature, oxygen) seem to be the most important factors determining the distribution of ostracods in the deep-sea, where some assemblages seem to be restricted to specific bathymetric ranges.

Analysis of ostracods communities of inner shelf and shallow marine settings allow to infer sea-level changes and coastal evolution, whereas deep-sea ostracods have been used to reconstruct bottom water-mass properties. Species distribution and morphological features of key species (i.e., size of the vestibule of *Krithe producta*) have been related with oxygenation, water depth and water-mass origin, whereas ostracodes Mg/Ca has been used successfully for reconstructing bottom-water temperatures.

6.3.8 Mollusks

Mollusca are one of the most diverse of all invertebrate Phyla. They are ubiquitous and its diversity is extreme, but only bivalves (Bivalvia) and pteropods (Gastropoda) have been used as paleoceanographic proxies.

Bivalves shell consists of two calcareous valves with growth lines. The majority of bivalves are marine, most are benthic living within or over the sediment-water interphase. Specific attributes of bivalve molluscs allow them to play an important role in marine ecosystems. They inhabit in brackish, saline and hypersaline transitional continental-marine environments, some are adapted to exposure and wave action in the intertidal zone, and at sea they are found from the shoreline down to abyssal depths.

Whereas temperature is recognized as the principal factor at large biogeographic scales, salinity is an important determining factor in the distribution of coastal and estuarine bivalves. The flow regime and the sediment composition are other physical parameters that seem to play a role in the spatial distribution of bivalves. Accordingly, bivalves have been extensively used as tools for paleoenvironmental reconstructions in coastal settings, where shells concentrations frequently occur allowing the recognition of sea-level changes and coastal evolution. Geochemical analyses have also been used to reconstruct sea-surface temperature (even at seasonal scale) by using oxygen isotopes, whereas salinity has been inferred based on $^{87}Sr/^{86}Sr$ ratios.

Pteropods (Gastropoda, Opistobranchiata) are ubiquitous marine faunas adapted to pelagic habitats limited to the narrow salinity ranges of oceanic water. These animals float or slowly sink in the water column, intermittently feeding and swimming. Some species possess a delicate external aragonite shell which is more soluble than calcite in seawater. Their distribution is controlled by temperature, food supply and oxygen. Accordingly, assemblages' composition reflects climatic and hydrological conditions. Pteropods have been recently proposed as a proxy for tracing aragonite dissolution in marine sediments and changes in carbonate chemistry of bottom waters.

6.4 Biomarkers as Paleoceanographic Proxies

The term biomarker includes molecular or chemical fossils, organic soluble compounds derived from distinctive biological molecules found in rocks and sediments. Most biomarkers are lipids, fatty and waxy compounds produced by a variety of organisms. After their biosynthesis and death of the precursor organism, biomarkers can survive deposition in sediments in a recognizable form. Thus, it may be possible to extract biomarkers from marine sediments to reveal evidences about the biological affinities of the precursor organisms and their habitat preferences.

The most common biomarkers used in paleoceanography are C_{37} alkenones and n-alkyl lipids.

6.4.1 Alkenones

The term alkenones is used to refer to a group of long-chain (C_{37}-C_{40}) straight methyl- and ethyl-ketones with one to four unsaturations produced by some phytoplankton species belonging to the Haptophycea and frequently preserved in sediments after deposition. Some of these species are coccolith producers. Haptophythe are very small and isolation of alkenones from individual species is not feasible. Thus, alkenones from sediments actually represent a composite contribution of all produced species present during deposition.

The unsaturated ratio of $C_{37:2}$, $C_{37:3,}$ and $C_{37:4}$ methyl-ketones is a reliable tool for past SST estimations. Two indices, which are linearly correlated with SST, have been proposed: the index Uk_{237} considers the concentration of $C_{37:2}$, $C_{37:3,}$ and $C_{37:4}$ methyl-ketones, whereas Uk'_{237} considers only the concentration of $C_{37:2}$ and $C_{37:3}$ methyl-ketones, since the proportion of $C_{37:4}$ compounds is low, especially in low- and mid-latitudes. Correlation between SST's and alkenone indices is usually higher than 0.9, although some alterations in the correspondence between them can be related with selective microbial degradation of alkenones types. Carbon and hydrogen isotopic measurements of alkenones are used to determine historical pCO_2 levels and sea-surface salinity (SSS) respectively.

6.4.2 N-Alkyl-Lipids

The bacteria (Archaea) are widespread distributed in the ocean, making up a large part of the pelagic biomass. An index based on the number of cyclopentane rings in glycerol dialkyl glycerol tetraethers (GDGT) found in sediments derived from archaea's cell membrane has been proposed to reconstruct SST. The GDGT's of the TEX86 index (TetraEther index of tetraethers consisting in 86 carbon atoms) is expected to derive from organisms belonging to the Crenarchaeota, which occurrence and diversity in the marine pelagic realm were noted in the 1990s. Not much is known on Crenarchaeota ecology and distribution, but the sedimentary signal of GDGT's seems to originate in the upper part of the water column. Calibration of the TEX86 from particulate organic matter in surface sediments correlates pretty well with mean temperature at the ocean surface.

6.5 Geological Proxies

Geological proxies are those non-biogenic elements contained in sediments that provide insights on paleoenvironmental changes, since their morphology, spatial distribution, sedimentary properties and mineralogical/geochemical content may have been controlled by environmental/climatic changes (Thomas and Burrough 2012; Thomas 2013). They are particularly useful when organic-based proxies are absent or not suitable for its study, as occurs in many continental settings (mainly drylands, Lyons et al. 2014). In the marine realm, although biological productivity is high and often well-preserved biological and biogeochemical proxies stand for the major sources of paleoceanographic, paleoclimatic and paleoenvironmental information, in some cases their poor preservation due to reworking, high-energy processes, erosion, diagenesis and other dynamic factors prevents their use as valuable tools for paleoreconstructions. Consequently, morphological, architectural and structural, as well as sedimentological and non-organic-derived geochemistry, can help in those reconstructions. Although as a general rule only sedimentological and geochemical proxies are considered as the exclusively really valuable tools for paleoreconstructions in the marine environment, some examples about the usefulness of large-scale bedforms and features (sedimentary structures and architecture) studied through bathymetric and seismic data demonstrate their capability to contribute to paleoreconstructions.

The broad spectrum of features—from large scale to discrete, individual components—offers a great variety of possibilities for a better understanding, interpretation and reconstruction of paleo-conditions at sea. As a result, the following categories of geological proxies and records are considered:

(1) Geomorphology
(2) Sedimentary architecture and structures
(3) Physical proxies
(4) Geochemical proxies.

6.5.1 Geomorphology

Geomorphological records (in this case better used than proxies) are considered as those features of the surface relief that can be interpreted as a function of their evolution. The term "geomorphological proxies" was used by Borgatti and Soldati (2010a, b) when discussing landslides (subaerial) as records for climate variability at a range of temporal and spatial scales. In the case of the submarine environments, geomorphological features relate to the sediment dynamics (erosional-depositional features) influenced by oceanographic factors, particularly the circulation of marine currents as well as the effect of gravity-downslope processes. The techniques that enable to evaluate the submarine relief and its evolution are bathymetric acoustic tools that depict bedforms and surface morphosedimentary features. Diverse geochronologic (dating) and isotopic tools on geological-sedimentological-biological-geochemical components contained in the geomorphological features can be suitable for obtaining absolute or relative ages. Submarine photography and video, and/or combination of all of them (as it is possible with the modern techniques deployed with ROV's) greatly help in the interpretation of sea-floor reliefs and present-day sedimentary processes.

Few examples are just herein mentioned where the authors apply the term "proxy" from bathymetric and seismic data. Lidz and Shinn (1991) used the morphology of Holocene shorelines and Pleistocene substratum as proxies for paleo sea-level reconstructions and timing of corals colonization. McAdoo et al. (2000) and Micallef et al. (2007) used submarine landslides morphology as a proxy for defining slope failures, their magnitude and variation through time, in the last case applied to the Storega mega-slide. McCave (2007) mentions the "large scale" approaches (sedimentary structures, acoustic profiles and seismic reflection) applied to the recognition of the importance of deep currents in the distribution of deep-sea sediments. Dunlop et al. (2011) used submarine glacial geomorphology in the Irish shelf as a proxy for deciphering ice-sheet processes, ice-flow dynamics and sediment pathways during deglaciation. Brothers et al. (2013) used multibeam bathymetry and morphology in a submarine canyon as proxies for determining the volume of sediments released during slope failures at the canyon walls. Dorschel et al. (2014) used high-resolution bathymetry together with geomorphological and biological data to define the geomorphological and sedimentological setting as well as sea-floor dynamics that condition the habitat of epibenthic faunas, defining erosive surfaces, escarpments and presence of ice-rafted debris as proxies for interpreting velocity and direction of bottom currents and timing of impact of scouring icebergs.

6.5.2 Sedimentary Architecture and Structures

These sedimentary records are in most of the cases useful for paleoreconstructions, because they contain different elements that enable to depict diverse aspects of the sedimentary construction.

Sedimentary architecture is the set of mega-features that compose the sedimentary blocks in a certain region, i.e., the sum of different sedimentary bodies. Therefore this is recognized at a regional scale. The main tool for interpreting sedimentary architecture is the seismic method. The seismic methods enable to reconstruct images of the subsoil up to different depths depending on the type of seismic method employed (high/low resolution, high/low frequency, shallow/deep penetration). Seismic-stratigraphy is the application of the sequence-stratigraphy concepts—originally established by Vail et al. (1977), Vail (1987) and presently updated by Catuneanu (2002, 2006)—to the geological interpretation of seismic records. Recognition of unconformities (geological discontinuities represented by conspicuous seismic reflectors) and its correlation with regional geological/climatic/oceanographic processes (sea-level transgressions and regressions, tectonic/volcanic events, gravity-downslope processes, erosion/deposition associated to ocean-currents dynamics, etc.) can be used as chronostratigraphic markers.

Sedimentary structures are macroscopic features of sediments that result from the depositional/erosional processes and were not substantially modified after deposition ("primary" sedimentary structures). Sedimentary structures range in size from millimeters or centimeters to large scale. For the larger structures the seismic method is useful for their recognition and interpretation, but for the smaller structures normally their preservation in sediment cores (gravity-, piston-, box-, multi-corers) is the only possible element in submarine research. Additionally, physical/geochemical properties of the sediments are useful proxies to interpret the timing and processes involved in the development of the sedimentary structures.

In both cases, sedimentary architecture and structures, their recognition and interpretation has been a powerful tool in paleoreconstructions at sea, as demonstrated in hundreds of examples given in the geological literature, particularly in the past decades for the determination of paleoceanographic processes involved in the evolution of contouritic drifts (e.g., Pickering et al. 1989, 1995; McCave 2007; Stow et al. 2002; Rebesco and Camerlenghi 2008).

6.5.3 *Physical Properties*

Physical properties are attributes of the bulk sediments constituted of inorganic elements, which result from the sum of processes occurred along the entire sedimentary cycle that involves weathering, erosion, transport and deposition along the pathway from the source areas to the sink place. In this way, certain properties such as grain-size, grain-size parameters, grain's shape and roundness, grain's surface textures, mineralogical composition, density and magnetic behavior are useful proxies for paleoprocesses and paleoenvironmental determinations. In many cases, postdepositional changes are reflected in the physical properties, which are therefore useful proxies for tracing diagenetic processes. McCave (2007) adequately synthesizes the basic concepts of the importance of physical properties as valuable proxies, particularly grain-size parameters of fine-grained and sandy sediments as indicators of flow speed involved in sediment transport and depositional conditions. Ledbetter (1986) had stressed the importance of using the silt fraction as the most valuable sediment fraction in flow speed studies and determination of critical shear stress for erosion and deposition. McCave et al. (1995) proposed the use of what they call "sortable silt" (particles size-fraction between 10 and 63 μm) which is considered as the grain-size interval more likely to be deposited individually in response to fluid stress. The rest of the physical properties, particularly magnetic parameters, have proved to be highly valuable in paleoreconstructions at sea as demonstrated in the enormous amount of publications on the matter. In this sense, St-Onge et al. (2007) give a detail of many centimeter-scale physical properties of sediments (magnetic susceptibility, gamma density, *p*-wave velocity, color reflectance and natural gamma radiation) and millimeter- to micrometer-scale core-imaging (W-ray, tomographies, X-ray, fluorescence, magnetic and nuclear magnetic resonance, scanning laser macroscopy and microscopy) and their importance as proxies for paleoceanographic studies.

6.5.4 *Geochemical Properties*

These properties are attributes of the elemental composition of sediments that were originally provided by the source rocks altered and modified along the exogenous sedimentary cycle and in the depositional environment as well. They probably are, together with the biological attributes, the most efficient proxies for paleoceanographic studies (Higginson 2009), which include past climate, changes in the properties of the water column, ocean circulation, sediment sources, sediment transport, marine organic production, postdepositional changes, among others (Calvert and Pedersen 2007). These authors provide the basic background and a short history of the application of geochemical proxies for paleoclimatic and paleoceanographic studies. The different chapters contained in Part 3 (Geochemical Tracers) of the book by Hilaire Marcel and de Vernal (2007) give most of the applications that these kinds of studies provide.

The interrelationship between lithogenous (inorganic components derived from rocks and soils of the continental crust), hydrogenous (inorganic phases precipitated from sea water) and biogenous components, provides elements for defining terrestrial versus marine deposition, source and provenance areas, eolian/fluvial/glacial inputs, terrestrial runoff, weathering regimes, precipitation and other climatically driven processes, terrestrial dust transport, diagenetic changes, biogeochemical processes in sea water, paleoproductivity, etc. Particularly, the ratio "element"/Al has been used as a proxy for determining terrestrial supply since Al is considered as the "lithogenous reference," and specifically Ti/Al ratio has proved to be highly valuable for that kind of studies. That ratio, as well as Si/Al, Zr/Al, K/Al, Cr/Al, Rb/Al, Zr/Rb, K/Rb, and others, provide information about terrestrial processes and conditions like aridity, runoff, precipitation, evaporation, weathering, etc. (see for example Calvert and Pedersen 2007). On the other hand, the relative abundance of certain elements (like Fe and Ti) also gives information about the input of terrigenous, siliciclastic sediments into the sea.

Another significant contribution to paleoreconstructions is provided by stable radiogenic isotopes like Pb, Nd, Sr, Os and Hf, which are respectively the end-products of the radioactive isotopes U-Th, Sm, Rb, Re and Lu. They provide information about terrestrial processes like weathering, erosion and transport of fluvial and eolian particles, as well as hydrothermal input, cosmic dust, dissolution of marine carbonates and paleo-bottom currents (Revel et al. 1996; Christelle and Hamelin 2007). The ratio between those elements (e.g., Sr/Nd, Sm/Nd, $^{143}Nd/^{144}Nd$) are also useful tools. Trace isotopes like Th and Pa, as the decay products of uranium isotopes dissolved in sea water, and its ratio Pa/Th, are valuable proxies for paleoflux and paleocirculation (Francois 2007).

References

Berger WH, Schulz M, Wefer G (2010). Quaternary oceans and climate change: lessons for the future? Int J Earth Sci (Geol Rundsch) 99(Suppl 1):S171–S189

Boltovskoy E (1981) Masas de agua en el Atlántico Sudoccidental. In: Boltovskoy D (ed) Atlas de Zooplancton del Atlántico Sudoccidental (Mar del Plata), pp 227–236

Borgatti L, Soldati M (2010a) Landslides and climatic changes. In: Alcántara-Ayala I, Goudie AS (eds) Geomorphological hazard and disaster prevention. Cambridge University Press, UK, pp 87–96

Borgatti L, Soldati M (2010b) Landslides as a geomorphological proxy for climate change: a record from the Dolomites (northern Italy). Geomorphology 120(1–2):56–64

Brothers DS, ten Brink US, Andrews BD, Chaytor JD, Twichell DC (2013) Geomorphic process fingerprints in submarine canyons. Marine Geology 337:53–66

Calvert SE, Pedersen TF (2007) Elemental proxies for paleoclimatic and paleoceanographic variability in marine sediments: interpretation and application. In: Hillaire-Marcel C, de Vernal A (eds) Proxies in Late Cenozoic Paleoceanography. Developments in Marine Geology 1(14), Elsevier, pp 567–644

Catuneanu O (2002) Sequence stratigraphy of clastic systems: concepts, merits, and pitfalls. J Afr Earth Sc 35(1):1–43

Catuneanu O (2006) Principles of sequence stratigraphy, 1st edn. Elsevier, 375 pp

Christelle C, Hamelin B (2007) Isotopic tracers of water masses and deep currents. In: Hillaire-Marcel C, de Vernal A (eds) Proxies in Late Cenozoic Paleoceanography. Developments in Marine Geology 1(15), Elsevier, pp 645–680

Dorschel B, Gutt J, Piepenburg D, Schröder M, Arndt JE (2014) The influence of the geomorphological and sedimentological settings on the distribution of epibenthic assemblages on a flat topped hill on the over-deepened shelf of the western Weddell Sea (Southern Ocean). Biogeosciences 11:3797–3817

Dunlop P, Sacchetti F, Benetti S, O'Cofaigh C (2011) Mapping Ireland's glaciated continental margin using marine geophysical data. In: Smith MJ, Paron P, Griffiths JS (eds) Geomorphological Mapping. Developments in Earth Surface Processes, vol 15. Elsevier, pp 339–354

Francois R (2007). Paleoflux and paleocirculation from sediment ^{230}Th and ^{231}Pa/^{230}Th. In: Hillaire-Marcel C, de Vernal A (eds) Proxies in Late Cenozoic Paleoceanography. Developments in Marine Geology 1(16), Elsevier, pp 681–716

Goudie AS (ed) (2004) Encyclopedia of Geomorphology. Routledge; International Association of Geomorphologists. London, New York, 1156 pp

Hay WH (2008) Evolving ideas about the Cretaceous climate and ocean circulation. Cretac Res 29:725–753

Higginson MJ (2009) Geochemical proxies (Non-Isotopic). In: Gornitz V (ed) Encyclopedia of paleoclimatology and ancient environments. Part of the series Encyclopedia of Earth Sciences Series. Springer, pp 341–354

Hillaire-Marcel C, de Vernal A (2007) Proxies in Late Cenozoic paleoceanography. Developments in Marine Geology 1. Elsevier, 843 pp

Ledbetter MT (1986) Bottom-current pathways in the Argentine basin revealed by mean silt particle size. Nature 321:423–425

Lidz BH, Shinn EA (1991) Paleoshorelines, reefs, and a rising sea: South Florida, U.S.A. J Coastal Res 7:203–229

Lyons R, Tooth S, Duller GAT (2014) Late Quaternary climatic changes revealed by luminescence dating, mineral magnetism and diffuse reflectance spectroscopy of river terrace paleosols: a new form of geoproxy data for the southern African interior. Quatern Sci Rev 95:43–59

McAdoo BG, Pratson LF, Orange DL (2000) Submarine landslide geomorphology, US continental slope. Mar Geol 169:103–136

McCave IN (2007) Deep-sea sediment deposits and properties controlled by currents. In Hillaire-Marcel C, de Vernal A (eds) Proxies in Late Cenozoic Paleoceanography. Developments in Marine Geology 1(1):19–62. Elsevier

McCave IN, Manighetti B, Beveridge NAS (1995) Circulation in the glacial North Atlantic inferred from grain-size measurements. Nature 374:149–152

Micallef A, Berndt C, Mason DG, Stow DAV (2007) Fractal statistics of the Storegga Slide. In: Lykousis V, Sakellariou D, Locat J (eds) Submarine mass movements and their consequences. Springer, The Nederlands, pp 3–10

Pickering KT, Hiscott RN, Hein FJ (1989) Deep-marine environments, clastic sedimentation and tectonics. Unwin Hyman, London, p 416

Pickering KT, Clark JD, Smith RDA, Hiscott RN, Ricci Lucchi F, Kenyon NH (1995) Architectural element analysis of turbidite systems, and selected topical problems for sand-prone deep-water systems. In: Pickering KT, Hiscott RN, Kenyon NH, Ricci Lucchi F, Smith RDA (eds) Atlas of deep-water environments: architectural style in turbidite systems. Chapman et al., London, pp 1–10

Rebesco M, Camerlenghi A (2008) Contourites. Elsevier, London, p 663

Revel M, Cremer M, Grousset FE, Labeyrie L (1996) Grain-size and Sr-Nd isotopes as tracer of paleo-bottom current strength, Northeast Atlantic Ocean. Mar Geol 131:233–249

St-Onge G, Mulder T, Francus P, Long B (2007). Continuous physical properties of cored marine sediments. In: Hillaire-Marcel C, de Vernal A (eds) Proxies in Late Cenozoic paleoceanography. Developments in Marine Geology 1(2):63–98. Elsevier

Stow DAV, Pudsey CJ, Howe JA, Faugères J-C, Viana AR (eds.) (2002) Deep-water contourite systems: modern drifts and ancient series, Seismic and Sedimentary characteristics. The Geological Society of London, Memoir 22, 466 pp

Thomas DSG (2013) Reconstructing paleoclimates and paleoenvironments in drylands: what can landform analysis contribute? Earth Surf Proc Land 38(1):3–16

Thomas DSG, Burrough SL (2012) Interpreting geoproxies of Late Quaternary climate change in African drylands: implications for understanding environmental change and early human behaviour. Quatern Int 253:5–17

Vail PR (1987) Seismic stratigraphy interpretation procedure. In: Bally AW (ed) Atlas of seismic stratigraphy: AAPG Studies in Geology 27(1):1–10

Vail PR, Todd RG, Sangree JB (1977) Seismic stratigraphy and global changes of sea level: Part 5. Chronostratigraphic significance of seismic reflections: Section 2. Application of seismic reflection configuration to stratigraphic interpretation. Memoir 26:99–116

Wefer G, Berger WH (1996) Global change and marine geology: introduction. Geol Rundsch 85:399–400

Chapter 7
State of the Art in the Paleoceanographic Reconstructions at the Argentina Continental Margin

Abstract The characteristics of the Argentina Continental Margin, particularly depending on its key location in the climate and oceanographic global system, make it to have a high potential for paleoclimatic, paleoceanographic and paleoenvironmental reconstructions. An overview of the present knowledge of the matter, given by the application of a varied set of proxies and tracers in different regions of the margin, like the shelf and the slope, is detailed.

Keywords Paleoreconstructions in the Argentine margin · Shelf · Slope-rise · Bioproxies · Sedimentological proxies · Geochemical proxies

For a better organization of this chapter, reconstructions in the shelf and slope/rise will be considered separately.

7.1 Reconstructions in the Continental Shelf

As a very extensive region with several large sedimentary basins, the Argentine continental shelf has potentially high sediment accumulation rates enabling high-resolution studies built on different records and proxies at varied scales (mega-macro- and micro-scales), and based on geological, biological and geochemical elements.

Bioproxies based on planktonic and shallow-water benthic faunas are among the most valuable tools for paleoreconstructions. This makes possible to precisely date climatic events in shallow marine records and to correlate them with continental data. The shelf records mainly the post-LGM transgression and luckily some previous (Quaternary) transgressive-regressive events, but studies are strongly limited by sediment reworking processes related with the landward migration of littoral systems during the transgressions and the ravinement process that accompanied the shoreline retreat. As a consequence, shelfal coastal evolution and related

R.A. Violante et al., *The Argentina Continental Margin*,
SpringerBriefs in Earth System Sciences, DOI 10.1007/978-3-319-04196-4_7

paleoceanographic changes are, in comparison to those from the adjacent coastal plains and offshore basins, relatively much less understood.

Most biological evidence from shelfal deposits comes from bivalves, but especially from foraminifers, ostracods and dinoflagellates.

From 1957 to 1960 the Lamont-Doherty Geological Observatory (presently named Lamont-Doherty Earth Observatory, LDEO) collected numerous piston cores (database at GeomapApp 2013) from the Argentine shelf containing shelly and conglomeratic shell layers underlying sandy facies (Urien and Ewing 1974; Guilderson et al. 2000). Malacological studies from some of those cores have been undertaken during the 1960s by Fray and Ewing (1963) and Richards and Craig (1963), and summarized in Richards (1966). Main results suggested pre-LGM shorelines at about 110 mwd (meters of water depth) between ~11,000 and ~17,250 years BP, but deeper than 120 mwd they were dated prior to 35,000 BP. However, it must be considered that this age of 35,000 years was at those times in the limit of the ^{14}C dating method possibilities, and that the time-frame of Quaternary sea-level fluctuations was not fully understood as it is today. Therefore, we presently definitively know that the date of 35,000 BP reflects the limit of the method but not the real age of the measured event. The molluscan species indicate a littoral to shallow neritic zone, and in general reflect waters colder than present. Thereafter, macrofossils from shelfal cores have received little attention. Recently, Bressan and Laprida (2014) have analyzed a shelly concentration from core SHN-T394 (40° 09′S–57° 14′W, 100 mwd) in the outer shelf off southeastern Buenos Aires province. They recognized composite concentrations, event deposits related with the Late Glacial. The shelly concentration was probably deposited in the lower shoreface-upper transition zone during storms.

The roots of foraminiferal micropaleontology in Argentina began in the mid-twenty century with Dr. Esteban Boltovskoy, who analyzed both benthic and planktonic foraminifera. He investigated numerous problems of local and global importance ranging from taxonomy and phylogeny to biogeography and water-masses. Many of his earlier papers were focused on ecology and the relation between salinity and foraminiferal distribution. Apart from documenting patterns of species distribution and dominance in restricted areas such as San Jorge Gulf and San Blas Bay (Boltovskoy 1954a, b), he also described brackish water and estuarine assemblages (Boltovskoy 1957, 1958; Boltovskoy and Lena 1971, 1974) and the most important species of foraminifers in the shelf areas of Argentina (Boltovskoy et al. 1980).

Additionally, Boltovskoy was a pioneer in the study of fossil foraminifera from South Atlantic sediment cores. He based his research mainly in the analysis of core samples from the Deep Sea Drilling Program (DSDP) and LDEO projects, focused particularly on biostratigraphic and paleoceanographic aspects. Boltovskoy (1973) analyzed 22 short cores (core length between ~1 and 8 m) from Buenos Aires shelf. Upper shelf cores (<50 mwd, core length between ~1 and 4 m) have retrieved exclusively Holocene sequences. Benthic foraminifers indicate coastal settings related with successive positions of the shoreline during the post-LGM transgression prior to 6000 BP. Middle shelf cores (50–100 mwd; core length between ~3 and 5 m) also contain littoral and inner shelf faunas (*Buccella peruviana*, *Elphidium discoidale*, *E. Ammonia becarii*, *Quinqueloculina seminulina*, *Bolivina striatula*, *Bulimina patagonica*, *Discorbis williamsoni*) probably associated with the beginning of the Holocene transgression. Very scarce planktonic fauna was found, especially *Neogloboquadrina pachyderma* and *G. glutinata*. Cold water species such as *Nonionella auris* and *Cassidulinoides parknerianus* were found at ~70 mwd indicating a very weak influence of the Malvinas Current on the shelf. Nowadays, the western boundary of the Malvinas Current lies along the 80–100 isobath (Boltovskoy et al. 1980). Cores form the outer shelf (>80 mwd; core length ~6 m) contain typical shelf fauna, a rather monothonous Holocene assemblages dominated by *B. peruviana*, *Q. seminulina* and *Epistominella exigua*. In the deeper cores, the Malvinas Current influence is evidenced by the presence of *N. auris* and *C. parknerianus*.

In the nearshore area of Salado Basin, Laprida et al. (2007) analyzed foraminifers from core SHN-T68 (36° 45.7′S–56° 37.2′W, 499 cm long. 12.80 mwd). The consistent dominance of the shelfal *B. frigida*/*Q. seminulum* group over the marginal marine, brackish-water *Elphidium* spp./*A. beccarii* group indicates normal marine to polyhaline conditions. Based on foram's assemblages, a relictic sandy tidal flat developed behind a barrier complex between 5400 and 4900 BP was recognized. Thereafter, between 4900 and 4300 BP, a smoothly sea-level rise has been detected associated with cold water invasion on the inner shelf. An amazing sterile clayey sequence ca. 1 m thick deposited at the same time after 4300 BP was interpreted as a deposit formed during a sea-level drop. A new pulse of sea-level rise ca. 3550 BP and the dominance of warm water on the inner shelf are suggesting global rather than regional forcing factors. Even when the reconstruction based on this facies description has wide indicative ranges regarding relative sea levels, the paleoenvironmental interpretation indicates that two pulses of fall in sea level have occurred between ~5800–5500 and ~3200–2800 BP, confirming a stepped model of Mid–Late Holocene sea-level fall in the north Argentine shelf.

Although some studies were made in combination with other (bio)proxies (i.e., Guerstein et al. 1992), palynological-based studies are still scarce, beginning in the early 2000s. Vilanova et al. (2008) analyzed dinocyst record of the above mentioned core SHN-T68 with respect to the distribution of modern dinocysts and other Holocene records from the adjacent continent. The low diversity and abundance of the cyst assemblages between ca. 5360 and 3300 ^{14}C yr BP can be related to a restricted littoral subtidal environments. After ca. 3300 ^{14}C yr BP a major change is characterized by an increase in dinocyst abundance, indicating a normal inner neritic environment. The dominance of *Operculodindium centrocarpum* at some levels suggests some influence of continental shelf waters whereas the higher abundances of dinocysts from heterotrophic taxa—e.g., *Protoperidinium stellatum*, *Votadinium calvum*, *V. spinosum*—in other levels reflects a more inshore-coastal water influence with increase nutrient availability. Present-day transitional coastal-neritic conditions were established after ca. 3300 ^{14}C yr BP.

In the Colorado Basin, marginal marine environments and inner shelf deposits have been identified at 50 mwd (core T1, 40° 30′S–60° 59′ 05″W, 560 cm; 45 mwd) by Bernasconi and Cusminsky (2007). *E. discoidale*, *B. peruviana campsi*, *Quinqueloculina patagonica* and *Quinqueloculina seminula* dominated the assemblages, indicating inner shelf environments. The sediments were deposited between ca. 9500 and 9000 BP, reflecting the rapid sea-level rise and high sedimentation rates during Early Holocene times in the area in response to the post-glacial transgression. Recently, Bernasconi and Cusminsky (2015) analyzed the distribution patterns of *Elphidium* aff. *poeyanum* and *B. peruviana* from the same core, identifying positive correlations between the abundance of *E.* aff. *poeyanum* and silt content and between *B. peruviana* and the sand content, reflecting better conditions for the development of both species. For the past 6500 years, Gómez et al. (2000, 2005) reconstructed sea-level changes investigating core PS2 (39° 07′ 24.11″S–61° 46′ 55.99″ O, 12.88 mwd) in the outer Bahía Blanca estuary. Sedimentological and micropaleontological aspects (palynology, ostracods and foraminifers) of the core have been studied previously by Gómez et al. (1992) and Guerstein et al. (1992). Foraminiferal assemblages indicate salt marsh deposits around 6350 BP, which evolved to an inner shelf environment, probably a tidal flat related to a sea-level rise and increased marine influence around 2300 BP, and finally in a high-energy coastal environment related to increasing sea level and strong tidal currents. Based on the same core, a negative (lower than present) sea-level oscillation was proposed thereafter by Gómez et al. (2006). According to the authors, the lower section was deposited immediately before the maximum Holocene transgression (6350 years BP) and consists of sediments representing a restricted intertidal environment, whereas the middle section represents an intertidal environment, probably a tidal flat in connection with tidal channel systems, indicating the occurrence of an important negative mean sea-level oscillation. The negative oscillation below the present sea level may be correlated with a worldwide climatic change that occurred around 2650 years BP.

More recently, Violante et al. (2014b) analyzed core SHN-T394 (40° 09′S–57° 14′W, 100 mwd) close to the shelf-slope border in the Colorado Basin, and recognized a sedimentary sequence in transgressive facies that records for the first time in the region the evolution of barriers-coastal lagoon environments during the last part of Marine Isotope Stage 2 close to the LGM. The sequence contains abundant calcareous microfossils, specially ostracods and benthic foraminifera. Microfossil associations and taphonomic signatures document the gradual transition from a fluvio-estuarine or brackish lagoonal pond developed in the foredune, with mixed reworked benthic foraminifera (*B. peruviana*, *Quinqueloculina* spp.) and brackish-water ostracoda (*Cyprideis salebrosa*, *Limnocytehre* spp.), to a tide-dominated shoreface related with the post-LGM transgression with typical Southwest South Atlantic inner shelf brackish fauna (*A. becarii*, *Elphidium gunteri*, *B. peruviana*, *Perissocytheridea whitensis*). The upper levels are dominated by outer shelf benthic assemblages with predominance of *Uvigerina striatula* and *Cassidulina inflata*.

For the Northern Patagonian shelf, Boltovskoy (1973) analyzed some cores ranging between 80 and 100 mwd. Foraminiferal assemblages indicate Early Holocene inner shelf deposits and limited Malvinas Current influence for the upper ~4 m of the sediment column. In core VM 12–36 (46° 04′S–65° 39′W, 82 mwd, 6 m long) (VEMA Cruise, Lamont-Doherty Earth Observatory, GeomapApp 2013), the lowermost samples represent pre-LGM times (>20.000 years BP), when an inner shelf environment (water depth 40–60 m) was established. LGM deposits are recognized between ~3–4 m below sea floor. Foraminifers indicate littoral to upper sublittoral environments related with the maximum Late Pleistocene low-stand. Thereafter, coastal and inner shelf environments are recognized in relation with the post-LGM transgression. No foraminifers were found in the upper 1.7 m from this core.

In the Southern Patagonian shelf, Boltovskoy (1954a) analyzed 9 short cores from San Jorge Gulf (water depth ~90 m; core length between ~0.20 and 1.40 m) finding only Holocene sediments with no outstanding vertical variations in faunal composition. However, total abundance varies and therefore this may be in relation with the Holocene sea-level history of the gulf. Foraminifers indicate inner shelf environments and cold subantarctic water dominance. More to the south, Bernasconi et al. (2009) analyzed core AU3C2 (Nuevo Gulf, water depth 143 m, core length 4.28 m) finding Late Holocene sediments containing benthic fauna. Qualitative and quantitative analyses confirmed dysoxic, marginal marine conditions probably related to a climate cooling which would have been the origin of water stagnation during the Little Ice Age.

Some studies were carried out in the South Malvinas Basin by Cusminsky (1992, 1994). In LGM times, a restricted (stratified?) environment was established associated with a ca. 150 m sea-level fall and the eastward deviation of the Malvinas Current. Some samples contained no foraminifers due to extreme dissolution, what indicates the advance of the polar front and the presence of carbonate undersaturated waters in the shelf. Overlaying Holocene sediments containing typical subantarctic foraminifera such as *N. pachyderma* and *Globigerina bulloides* reveal the restoration of the northward-flowing cold Malvinas Current.

Exclusively ***geological* (*non-biogenic*)** mega-, macro- and micro-proxies and records have been used in the shelf for determining sea-level fluctuations, sediment sources and coastal/nearshore sedimentary processes. Sedimentary sequences preserved in cores, together with the broad physical properties of sediments (in some cases associated to microfaunistic studies already mentioned), were performed by Fray and Ewing (1963), Richards and Craig (1963), Parker et al. (1978, 2008), Urien and Ewing (1974), Parker and Violante (1982), Guilderson et al. (2000), Laprida et al. (2007) and Violante et al. (2014a). These studies, considered in the framework of the sedimentary architecture known from seismic-stratigraphic analyses (Ewing and Lonardi 1971; Urien and Ewing 1974; Violante and Parker 2000, 2004; Violante 2003; Parker et al. 2008) allowed to depict the sequence of Quaternary transgressive-regressive events associated to the alternation of interglacial/glacial periods. Violante and Parker (1993) and Parker et al. (2008)

identified four sequences of that age (Depositional Sequences 4 to 1 from base to top, locally known respectively as Interensenadense, Belgranense, San Clemente Fm. and Holocene transgression—this one known under different formational names).

Physical properties of sediments have been traditionally used as proxies for sediment composition and source areas. Since as early as in the mid 20th century, diverse authors that studied coastal and nearshore sediments through their mineralogical composition (Teruggi 1954; Etchichuri and Remiro 1960, 1963; Depetris and Griffin 1968; Bercowsky 1986; Gelos et al. 1988; Potter 1994; Campos et al. 2008; see more details in Violante et al. 2014c) determined that they are dominantly terrigenous, sourced in two main regions: (1) the Patagonian Andes and plains (volcanic-pyroclastic mineralogical assembly belonging to the Pampean-Patagonian association), and (2) the Brazilian shield (igneous-metamorphic mineralogical assembly). The first one is dominant in the entire coasts and shelf and the second one in the area adjacent to the Rio de la Plata. The processes that intervened in the sediment transfer from the continent to the sea from the Patagonian region have been interpreted as being 56% from coastal erosion, 41% from eolian transport and 3% from fluvial transport (Gaiero et al. 2003; Violante et al. 2014c), with a total amount of sediments transferred to the sea bypassing the Patagonian coasts of 70 Mt a year (Pierce and Siegel 1979; Gaiero et al. 2003). Regarding coastal erosion, it must be considered that the sedimentary sequences exposed in the coast, which suffer intense erosion by marine processes, are composed of sediments belonging to the Pampean-Patagonian mineralogical association (Teruggi 1954) that later feeds the shelf. In the case of the Rio de la Plata region, sediment supply has been considered to range between 57 and 130 Mt a year (Depetris and Griffin 1968; Depetris and Paolini 1991; Giberto et al. 2004; Campos et al. 2008). Other authors used clay mineralogy to analyze sediment sources and transport in the shelf (Biscaye 1965; Griffin et al. 1968; Depetris 1996; Gaiero et al. 2002; Nagai et al. 2013).

The first geochemical studies with implications on the composition of marine sediments were done in continental areas close to the Atlantic coast (Gaiero et al. 2002, 2003, 2007; Depetris et al. 2005), where diverse terrigenous elements (Si, Al, Mg, Ca, Mn, Ti, Na, K, Sr Ba, Sc, Y, Rb, Sr, Nd), heavy metals (Fe, Pb, Cu, Ni, Cr, Zn, Co) and rare earths (Zr, Th, La, Ce, Pr, Nd, Sm, Eu, Gd, Tb, Dy, Ho, Er, Tm, Yb, Lu, in some cases in combination with biogeochemical components) were measured from outcrops as well as from fluvial and eolian deposits, resulting in the determination of sediments source areas and fluvial/eolian transport toward the adjacent shelf. Differences in sediment supply between glacial and interglacial times, as well as the importance of volcanic eruptions as producers of sediments, were determined. Furthermore, Mahiques et al. (2008) stated, by measuring Nd and Pb isotopic compositions in surface samples from the shelf, the influence of subantarctic shelf waters along the Argentine margin, with a significant input from the Rio de la Plata in the northernmost part of the shelf and in Uruguayan a Brazilian waters as well.

7.2 Reconstructions in the Slope and Rise

The Argentine continental slope, given its importance as an archive for paleoceanographic and paleoclimatic reconstructions—a lot more than what the shelf can provide—has been a field of study for the application of a numerous kind of proxies. Therefore, for a better description of the achieved results it is useful to try each kind of proxies as a separate item.

7.2.1 Paleoceanographic and Biostratigraphic Background Offshore the Argentina Continental Margin

The western South Atlantic has been object of several geological, geochemical and micropaleontological studies, many of them performed in the Argentine continental slope and offshore (e.g., Boltovskoy 1962, 1966, 1970, 1981; Boltovskoy et al. 1996; Chiessi et al. 2007, 2014, 2015; Hernández-Molina et al. 2009, 2010, 2015; Toledo et al. 2007, 2008; Violante et al. 2010; Laprida et al. 2011; Bozzano et al. 2011; Pivel et al. 2013; Preu et al. 2012, 2013; García Chapori et al. 2015; Voigt et al. 2013, 2015). The Austral and Colorado Basins concentrate mostly biostratigraphic published and unpublished studies performed in cores, recording up to Early Cretaceous sediments. These were based on the analysis of nannoplancton, nannoconids, silicoflagelates, calpionelids and benthic foraminifera (e.g., Malumián 1970; Malumián and Náñez 1996; Pérez Panera and Angelozzi 2006; Pérez Panera et al. 2006; Malumián and Náñez 1991; Caramés and Malumián 2006; Pérez Panera 2012, 2013, 2015). In the Malvinas Plateau and the western flank of the Mid-Atlantic Ridge (Argentine Basin), several sites have been drilled by the Deep Sea Drilling Project and recorded relatively complete sequences of Mesozoic and Cenozoic sediments. Upper Jurassic and Lower to Upper Cretaceous sediments were delineated within the Mesozoic sequences of the western edge of the M. Ewing Bank; whereas Paleocene–Eocene, Middle Eocene, Upper Eocene–Oligocene, Middle-Upper Miocene and Pliocene-Quaternary sediments were recovered on the Malvinas Plateau. Within this context, several groups of fauna and flora (foraminifera, radiolaria, pelecypods, ammonites, belemnites, sponges, nannoplankton, diatoms, silicoflagellates, dinoflagellates, calcisphaerulids, spores and pollen) were analyzed in order to achieve a regional stratigraphy (Basov et al. 1983). This report included a description of the species present in the region, and biostratigraphic and paleoenvironmental analyses of all the analyzed sequences (e.g., Basov and Krasheninnikov 1983; Gombos and Ciesielski 1977; Harris 1977; Tjalsma 1977; Wind and Wise 1983; Wise 1983; among others). The first attempt at such a reconstruction in the Malvinas Plateau was the one made by Tjalsma (1977) who used foraminiferal species diversity for estimating relative temperature changes. After that, and taking into account Leg 36 and Leg 71 data, Krasheninnikov and Basov (1983) performed a temperature curve for the Malvinas Plateau and the western flank of the Mid-Atlantic Ridge for the Cenozoic. In it, the authors pointed out: (a) a drop in temperature during the Mesozoic/Cenozoic boundary; (b) relatively warm climatic conditions for the Late Paleocene-Early Eocene; (c) cold climatic conditions during the Late Miocene; (d) an irreversible climatic deterioration established in the Late Pliocene which corresponded to a northward migration of the Polar Front ending in a consequent rise in the level of the Calcium Compensation Depth and the absence of calcareous fauna in the Upper Pliocene–Pleistocene sediments.

Most studies dealing with the Quaternary paleoceanographic reconstruction of the Argentina Continental Margin are based on sedimentological and geochemical proxies. Even though, micropaleontological analyses have been a significant tool in the advance of this discipline. Boltovskoy (1962, 1966, 1973, 1979) was the first one in performing (qualitative) paleoceanographic inferences based on the analysis of drilling sites data of the ACM. In these contributions, the author established (qualitatively) the latitudinal position of the Brazil–Malvinas Confluence from the presence/absence of specific species of planktonic foraminifera and their abundance, and also from the morphology predominance of *N. pachyderma*. Some time later, he linked the regional paleoceanographic changes recognized in the ACM with the global paleogeographic events that had taken place since the Miocene. Within, he proposed paleoclimatic curves for the Pliocene-Holocene period and recognized some of the classical glacial–interglacial cycles of the European chronology (Würmiense—Rissiense). Paleoceanographic quantitative reconstructions were possible only after the development of multivariate statistical analyses. This allowed obtaining numerical values of sea-surface temperature for this sector of the western South Atlantic. Between the calibrations performed in the ACM are: (a) the analyses of the oxygen isotopic composition of *Globorotalia inflata* and *Globorotalia truncatulinoides* as trackers of the latitudinal shifts in the BMC; (b) sea-surface temperature Mg/Ca ratios of *G. inflata*; and (c) sea-surface temperature calibration function F141-35-5 based on planktonic foraminfera (Chiessi et al. 2007; Groeneveld and Chiessi; 2011; García Chapori et al. 2015). However, later publications proposed more accurate sea-surface temperature calibration techniques that included morphogenetic analysis of different species of planktonic foraminifera. The aim of these studies was to detect new 'types' within each of these species, and assess their biogeographic patterns and the link these have with the surface-ocean properties (Morard et al. 2011, 2015; Quillévéré et al. 2011). For the northern sector of the ACM, the new quantitative reconstructions evidenced variations in $\delta^{18}O$ values of *G. inflata* and anomalies in the sea-surface temperature patterns, revealing orbital-to-millennial scale migrations of the Brazil–Malvinas Confluence between the Mid Pleistocene and Early Holocene as a response to changes in the position and strength of the northern portion of the South Western Winds. After that, the oceanic conditions would have been stable since Mid-Holocene until the present (Laprida et al. 2011; García Chapori et al. 2015; Voigt et al. 2015). This is in agreement with paleoceanographic reconstructions performed in the Brazil Margin for the same period (Pivel et al. 2013; Chiessi et al. 2014, 2015). Qualitative and quantitative analyses have also been applied to deep foraminiferal assemblages in order to reconstruct variations in the vertical structure of the water-masses and their evolution during the Upper Neogene. However, most of them are not exclusively from the Argentina Continental Margin (Lohmann 1978; Boltovskoy et al. 1980; Schnitker 1980; Boltovskoy 1981; Boltovskoy and Totah 1987, 1988; Johnson 1983; Hodell et al. 1983; Harloff and Mackensen 1996). For instance, some authors recognized hiatuses and dissolution cycles throughout the Neogene on the basis of benthic foraminifera analyses (Ciesielski and Wise 1977; Ciesielski et al. 1982; Bertels 1984a, b; Bertels and Núñez 1989; Cusminsky 1992, 1994), related to the activity of the AABW and variations in the depth of the lysocline during Pleistocene cold periods (Boltovskoy 1973). Latest results from factor analysis evidenced that benthic foraminiferal assemblages of the Argentina continental slope were mainly composed of high organic matter and oxygen availability-associated species revealing that the productivity signal has been the main factor in determining the structure of the benthic communities during the Mid-Late Pleistocene. The resulting analyses reflected a change from low to high productivity, characterized by a strong seasonal regime. These variations in the exported productivity would be related to changes in the shelf-break upwelling of Patagonia (García Chapori et al. 2014; Laprida et al. 2014).

7.2.2 Geomorphological Records

The present-day morphosedimentary features that shape the ACM are the result of a complex history of evolution under a multiple set of variables that include climatic and oceanographic changes, sea-level fluctuations, glacioisostasy, tectonic and varied sedimentary dynamics encompassing along- and down-slope processes. Features like terraces, ridges, drifts, scarps, channels, canyons, lobes, etc., are the most conspicuous components of the relief, which contain elements that can be considered as proxies for paleoclimatic, paleoceanographic and paleosedimentary processes. Following Borgatti and Soldati (2010a, b), geomorphological records are herein considered as valuable tools with implications in sedimentary processes and climate variability.

Detailed morphological studies in the ACM are still lacking in many regions. However, valuable research was locally done on the basis of: (a) multibeam surveys (e.g.,: Krastel et al. 2011; Franco Fraguas et al. 2014 in the offshore of Rio de la Plata; Muñoz et al. 2012 in a sector the Patagonian margin; von Lom Keil et al. 2002 in the Argentine basin); (b) morphological reconstructions based on seismic surveys at both, regional and local levels (e.g.,: Ewing and Lonardi 1971; Ewing et al. 1973; Urien and Ewing 1974; Hernández-Molina et al. 2009, 2010; Preu et al. 2012, 2013); and (c) bathymetric analysis (Costa et al. 2014; Violante et al. 2014a, c). Although a lot of work has yet to be done, the mentioned studies provide insights on the recent evolution of the slope at a regional scale, what in combination with other physical and geochemical proxies helped to better understand the timing of different evolutionary processes and some paleoenvironmental conditions in the region.

7.2.3 Sedimentary Architecture and Structures as Records

The interpretation of the sedimentary architecture of the ACM based on the analysis of seismic records is one of the most important elements suitable for providing evidences and timing of the regional evolution, sedimentary processes and oceanographic changes. Ewing and Lonardi (1971), Lonardi and Ewing (1971), Ewing et al. (1973), Klauss and Ledbetter (1988), Urien and Zambrano (1996), Hinz et al. (1999) and Franke et al. (2007) settled the basis of the chronology of the events shaping the margin, as well as the knowledge of the large-scale sedimentary processes, when the most prominent seismic reflectors recognized in seismic records were correlated with major global and regional climatic and/or oceanographic changes. Ortiz Jaureguizar and Cladera (2006) and Cavallotto et al. (2011) synthesized the major tectonic, climatic and oceanographic conditioning factors involved in the evolution of the ACM. Later on, more detailed interpretation of seismic reflectors and associated seismic sequences (both at a local and regional level) led to depicting specific processes involved in the margin construction and evolution, like the formation and development of contouritic drifts, turbiditic deposits, submarine canyons, etc. (e.g.,: Parker et al. 2008; Hernández-Molina et al. 2009, 2010; Violante et al. 2010; Bozzano et al. 2011; Krastel et al. 2011; Lastras et al. 2011; Gruetzner et al. 2011, 2012, 2016; Preu et al. 2012, 2013; Voigt et al. 2013, 2016).

As a result of the analysis of the sedimentary architecture and megastructures, including the limiting unconformities represented by seismic horizons separating the major sedimentary sequences, diverse key-events in the evolution of the Argentine margin were defined (Table 3.2), such as: (a) the development of impressive continental volcanic/magmatic episodes associated to the opening of the South Atlantic, seismically manifested as strong seaward-dipping reflectors; (b) the big expansion of the Antarctic ice-sheets that significatively affected the thermohaline circulation and induced modifications in the Argentine margin configuration at the Eocene–Oligocene boundary and later on in the mid-Miocene; (c) the first development of contouritic drifts since the Eocene–Oligocene as a result of the installation of the present, definitive oceanographic configuration of the Southwestern Atlantic Ocean; (d) the formation of the Ewing Terrace in the mid-Miocene; (e) the activation/reactivation of the strong marine currents (AABW, CDW, NADW, AIBW) in different times of the margin evolution, in most of the cases causing the increasing in erosive/depositional processes that produced important unconformities represented by seismic horizons; (f) the beginning of the high-frequency sea-level changes at the Tertiary/Quaternary boundary.

7.2.4 Physical Proxies

The study of the sediment's physical properties in the slope and rise leads to the interpretation of the sediment dynamics in the context of the Source-to-Sink concept, therefore providing insights also on the sediment source areas. The first indicators about sedimentary processes associated to ocean currents was provided by Ewing and Lonardi (1971), and then by Biscaye (1965) and Griffin et al. (1968), who proposed the marine transport and distribution of clays along the Argentine margin from south to north by the activity of the AABW. Ledbetter (1986, 1993) provided information about bottom current's relative velocities and sediments pathways. Physical properties such as grain-size and other sedimentological parameters have been used to determine sea-level fluctuations and sedimentary facies (Violante et al. 2010; Bozzano et al. 2011). Although not related to physical properties of sediments but to land-derived biological content studied in cores with the aim of evaluating sedimentation rates, Groot et al. (1965) determined that sedimentation rates during Late glacial times in the Argentine Basin was between 2 (?) and 13.8 cm/103 years, what decreased to 0–5.8 cm/103 years in post-glacial times; the largest sediment supply to the basin in glacial times was considered by the authors to be the consequence of the increasing in the area of exposure of the shelf to subaerial conditions during low-stands, together with stronger winds and currents, what provoked greater erosion and larger seaward sediment transport. In recent years, physical properties, in some cases combined with ^{14}C datings, have been used in more detail to depict sedimentary dynamics, along- and across-slope sedimentary processes and slopes stability (Violante et al. 2010; Bozzano et al. 2011; Krastel et al. 2011; Ai et al. 2014). Recently, significant advances have been made in more specific subjects such as the investigation of carbonate content as a contributor to the evolution of planktonic faunas (Baumann et al. 2004), the use of X-ray mineral analysis, magnetic mineralogy and magnetic susceptibility to evaluate sedimentation, mineral alteration and diagenetic changes (Garming et al. 2005; Riedinger et al. 2005), the application of grain-size and elemental ratios for determining terrigenous versus carbonatic components and its importance for defining sediment sources and sedimentary processes (Voigt et al. 2013), the combination of sedimentological parameters with hydrological and productivity proxies to evaluate sedimentary processes and direction of movement of currents (Franco Fraguas et al. 2014, 2016), and the sedimentary differences useful for determining fluctuations in the water-masses boundaries, sediment dynamics and supply of organic carbon and carbonates (Frenz et al. 2004).

7.2.5 Geochemical Proxies

The application of geochemical proxies from lithogenous components of sediments in the slope is a very recent matter. Garming et al. (2005) and Riedinger et al. (2005) established, on the basis of geochemical as well as pore-water and solid-phase analyses, variations in sedimentation rates and diagenetic processes during glacial and interglacial times, and in particular the second mentioned paper demonstrated, for the northernmost part of the ACM, a drastic decreasing in sedimentation rate from a few hundred to only 5 cm/kyr during the Pleistocene–Holocene transition. Riedinger et al. (2014) used the same methods to evaluate biogeochemical changes at the seafloor.

Information about sediment sources was provided by several authors from diverse points of view. Noble et al. (2012), working in the Scotia sea, used ^{230}Th normalized fluxes and isotopes of Pb, Nd, and Sr to evaluate source areas, transport processes, and oceanographic controls, concluding that continental detrital fluxes during glacial times were significatively higher than in the Holocene, added to the fact that marine currents transport was also higher at that time due to the increased speeds of the Antarctic Circumpolar Current. In terms of oceanographic control, Bender et al. (2013) determined in the northernmost region of the slope, by using the Ti/Ca ratio in relation to grain-size, the shifting of the subtropical Shelf Front from 36°S to the north (and hence the northward expansion of the Malvinas Current) at the beginning of the Holocene as a result of climate changes associated to variations in the latitudinal position of the southern westerly winds. Mollenhauer et al. (2006), based on ^{230}Th from top cores together with other biological proxies (foraminifera and alkenones), call the attention on the lateral displacement of suspended particles in the water column due to currents circulation prior to deposition, what is very important for evaluating the real location of bottom samples (both sedimentary and biogenic) when they were moved laterally from their point of origin in the sea surface and are used for past environmental reconstructions. Voigt et al. (2013) used grain-size, RX, XRF scanning and bulk geochemistry of major and minor elements, in combination with biological proxies, to determine source areas (using the Ca/Fe and Si/Al ratios) and sedimentary processes associated to the dynamics of the Mar del Plata Canyon and surroundings. Razik et al. (2015) applied cluster analytical methods to enviromagnetic, major element and grain-size data to evaluate sources and transport characteristics of shelf and slope sediments of a large area of the South Atlantic that includes the ACM, defining three major sediment-source areas with different petrographic and climatic characteristics, and their implications in the sea-floor sediment deposition. Franco-Fraguas et al. (2016) in the Uruguay continental margin—therefore with implicances in the northernmost part of the Argentine margin,—proxies such as Nd., Al/Si and Fe/K (lithogenic) and C/N, δ^{13}C and δ^{15}N (organic matter) and elemental compositions such as Al, Ti, Fe, K, Mn, Ca, P, Mg, Ba and Si were used for defining sediment provenance, sedimentary and organic matter contribution of the Rio de la Plata to the slope, surface productivity and environmental control. In a regional rather than local work which comprises a large part of the Atlantic Ocean that includes the Argentine margin, Govin et al. (2012) measured major elements (Ca, Fe, Al, Si, Ti, K and Ca) in combination with elemental ratios (Ti/Al, Al/Si, Fe/K, Fe/Ca, Ti/Ca) to reconstruct sediment sources and paleoclimatatic conditions (humid/dry climates, physical/chemical weathering).

Finally, and in relation to the study of sea-floor instabilities, Henkel et al. (2011) used, in an area of the slope in front of the Rio de la Plata, geochemical analyses of pore water and sulfate determinations, together with some biogenic components (organic carbon), supported by sedimentological and geotechnical analyses (undrained shear stress, moisture and density parameters), to define paleosurfaces in the sedimentary sequences that could originate mass transport processes and slides, identifying a particular gravity flow probably associated to an earthquake occurred in 1988 in the subsoil of the Río de la Plata.

References

Ai F, Strasser M, Preu B, Hanebuth TJJ, Krastel S, Kopf A (2014) New constraints on oceanographic vs. seismic control on submarine landslide initiation: a geotechnical approach off Uruguay and northern Argentina. Geo-Mar Lett 34:399–417

Argentina Hydrographic Survey (2015) Cores and bottom samples database

Basov IA, Krasheninnikov VA (1983) Benthic foraminifers in Mesozoic and Cenozoic sediments of the Southwestern Atlantic as an indicator of paleoenvironment. Init Rep Deep Sea Drilling Proj 71:739–787, Washington (U.S. Government Printing Office)

Basov IA, Ciesielski PF, Krasheninnikov VA, Weaver FM, Wise SW Jr (1983) Biostratigraohic and Paleonotologix synthesis: Deep Sea Drilling Project Leg 71, Falkland Plateau and Argentine Basin. Init Rep Deep Sea Drilling Proj LXXI:445–460, Washington (U.S. Government Printing Office)

Baumann K-H, Böckel B, Donner B, Gerhardt S, Henrich R, Vink A, Volbers A, Willems H, Zonneveld KAF (2004) Contribution of calcareous plankton groups to the carbonate budget of South Atlantic surface sediments. In: Wefer G, Mulitza S, Ratmeyer V (eds) The South Atlantic in the Late Quaternary. Springer, pp 81–99

Bender VB, Hanebuth TJJ, Chiessi CM (2013) Holocene shifts of the Subtropical Shelf Front off southeastern South America controlled by high and low latitude atmospheric forcings. Paleoceanography 28:1–10

Berkowsky F (1986) Arenas del Río de la Plata: una excepción a la relación entre composición de areniscas y la tectónica de placas. In: Primera Reunión Argentina de Sedimentología, La Plata. Asociación Argentina de Sedimentología. Actas de Resúmenes: 263–266

Bernasconi E, Cusminsky GC (2007) Foraminíferos bentónicos de un testigo holocénico de la plataforma continental argentina (40° 30′ 00″S y 60° 59′ 05″O). Ameghiniana 44:271–278

Bernasconi E, Cusminsky GC (2015) Study of the distribution of *Elphidium* aff. *poeyanum* (D'ORBIGNY) and *Buccella peruviana* (D'ORBIGNY) from the Colorado basin (South America): Holocene paleoenvironmental inferences. The Holocene. doi:10.1177/0959683615571424

Bernasconi E, Cusminsky G, Gómez EA (2009) Foraminíferos bentónicos del Holoceno del Golfo Nuevo, Argentina: inferencias paleoclimáticas. Rev Esp Micropaleontol 41:21–34

Bertels A (1984a) Ciclos de disolución de foraminíferos planctónicos en el océano Atlántico Sudoccidental Austral: significancia paleoceanográfica durante el Mioceno superior-Pleistoceno inferior. III Congreso Latinoamericano de Paleontología: Memoria: 524–538

Bertels A (1984b) Foraminíferos planctónicos del Neógeno del Océano Atlántico Sudoccidental Austral. Rev Esp Micropaleontol 17(2):221–252

Bertels A, Nuñez HJ (1989) Micropaleontología y paleomagnetismo de sedimentos del Océano Atlántico Sudoccidental Austral en el Neogeno: sus hiatos. Rev Esp Micropaleontol 21 (3):391–408

Biscaye PE (1965) Mineralogy and sedimentation of recent deep-sea clay in the Atlantic Ocean and adjacent seas and oceans. Geol Soc Am Bull 76:803–832

Boltovskoy E (1954a) Foraminiferos del golfo San Jorge. Inst. Nac. Invest. Cienc. Nat Rev Geol 3 (3):85–246

Boltovskoy E (1954b) Foraminiferos de la bahia San Blas. Inst. Nac. Invest. Cienc. Nat Rev Geol 3(4):247–300

Boltovskoy E (1957) Los foraminiferos del estuario del Rio de la Plata y su zona de influencia. Inst. Nac. Cienc. Nat Rev Geol 6(1):1–76

Boltovskoy E (1958) The foraminifera fauna of the Rio de la Plata and its relation to the Caribbean area: Cushman Foundation of Foraminiferal Research. Contribution 9(1):17–21

Boltovskoy E (1962) Planktonic foraminifera as indicators of different water masses in the South Atlantic. Micropaleontology 8(3):403–408

Boltovskoy E (1966) La zona de convergencia Subtropical/Subantártica en el Océano Atlántico (Parte occidental). Secretaria de Marina, Servicio de Hidrografia Naval, H. 1018:1–37

Boltovskoy E (1970) Masas de agua (característica, distribución, movimientos) en la superficie del Atlántico sudoeste, según indicadores biológicos—foraminíferos, Argentina. Servicio de Hidrografia Naval H-643:1–99

Boltovskoy E (1973) Estudio de testigos submarinos del Atlántico Sudoccidental. Revista del Instituto Nacional de Investigación Natural y Museo Argentino de Ciencias Naturales "Bernardino Rivadavia", Ciencias Geológicas 7(4):215–240

Boltovskoy E (1979) Paleoceanografia del Atlántico Sudoccidental desde el Mioceno, según estudios foraminiferológicos. Ameghiniana 16:357–389

Boltovskoy E (1981) Masas de agua en el Atlántico Sudoccidental. In: Boltovskoy D (ed) Atlas de Zooplancton del Atlántico Sudoccidental (Mar del Plata): 227–236

Boltovskoy E, Lena H (1971) The foraminifera (except family Allogromiidae) which dwell in fresh water. J Foram Res 1(2):71–76

Boltovskoy E, Lena H (1974) Foraminiferos del Rio de la Plata, Argentina. Servicio de Hidrografía Nava, H-661:1–22

Boltovskoy E, Totah VI (1987) Relación entre masas de agua y foraminíferos bentónicos en el Pacífico Sudoriental. Physis 45(109):37–46

Boltovskoy E, Totah VI (1988) Masas de agua y foraminíferos bentónicos en el cono sur de América Meridional. Rev Esp Micropaleontol 20(3):445–450

Boltovskoy E, Giussani M, Watanabe S, Wright R (1980) Atlas of benthic shelf foraminifera of the southwest Atlantic. W. Junk, The Hague, 147 pp

Boltovskoy E, Boltovskoy D, Correa N, Brandini F (1996) Planktic foraminifera from the southwestern Atlantic (30–60 S): species-specific patterns in the upper 50 m. Mar Micropaleontol 28(1):53–72

Borgatti L, Soldati M (2010a) Landslides and climatic changes. In: Alcántara-Ayala I, Goudie AS (eds) Geomorphological hazard and disaster prevention. Cambridge University Press, pp 87–96

Borgatti L, Soldati M (2010b) Landslides as a geomorphological proxy for climate change: a record from the Dolomites (northern Italy). Geomorphology 120(1–2):56–64

Bozzano G, Violante RA, Cerredo ME (2011) Middle slope contourite deposits and associated sedimentary facies of NE Argentina. Geo-Mar Lett 31:495–507

Bressan GS, Laprida C (2014) Depósitos de evento del Pleistoceno Superior en la plataforma externa, sector norte del Margen Continental Argentino. Actas 19° Congreso Geológico Argentino, Córdoba, Argentina

Campos EJD, Piola AR, Matano RP, Miller JL (2008) PLATA: a synoptic characterization of the southwest Atlantic shelf under influence of the Plata River and Patos Lagoon outflows. Cont Shelf Res 28:1551–1555

Caramés A, Malumian N (2006) Foraminíferos bentónicos del Cenozoico del pozo Estrella x-l, Cuenca del Colorado, plataforma continental argentina. Ameghiniana 37(4):387–419

Cavallotto JL, Violante RA, Hernández-Molina FJ (2011) Geological aspects and evolution of the Patagonian continental margin. Biol J Linn Soc 103:346–362

Chiessi CM, Ulrich S, Mulitza S, Pätzold J, Wefer G (2007) Signature of the Brazil-Malvinas Confluence (Argentine Basin) in the isotopic composition of planktonic foraminifera from surface sediments. Mar Micropaleontol 64(1):52–66

Chiessi CM, Mulitza S, Groeneveld J, Silva JB, Campos MC, Gurgel MH (2014) Variability of the Brazil Current during the late Holocene. Palaeogeogr Palaeoclimatol Palaeoecol 415:28–36

Chiessi CM, Mulitza S, Mollenhauer G, Silva JB, Groeneveld J, Prange M (2015) Thermal evolution of the western South Atlantic and the adjacent continent during Termination 1. Clim Past 11:915–929

Ciesielski PF, Wise SW (1977) Geologic history of the Maurice Ewing Bank of the Falkland Plateau (southwest Atlantic sector of the Southern Ocean) based on piston and drill cores. Mar Geol 25(1):175–207

Ciesielski PF, Ledbetter MT, Ellwood BB (1982) The development of Antarctic glaciation and the Neogene paleoenvironment of the Maurice Ewing Bank. Mar Geol 46:1–51

Costa IP, Violante RA, Paterlini CM, Schwenk T, Bozzano G (2014) Deslizamientos submarinos en el talud continental adyacente al Sur bonaerense. XIX Argentina Geological Congress, Córdoba, Argentina. Abstracts volume S18-4. Asociación Geológica Argentina, ISBN: 978-987-22403-5-6

Cusminsky GC (1992) Foraminíferos Bentónicos provenientes de testigos del Océano Atlántico Sudoccidental Austral. Rev Esp Micropaleontol 24(1):5–32

Cusminsky GC (1994) Estudio micropaleontológico (foraminíferos) de dos testigos provenientes del Océano Atlántico Sudoccidental Austral. Rev Esp Micropaleontol 26(2):109–123

Depetris PJ (1996) Riverine transfer of particulate matter to ocean systems. In: Ittekkot V, Schäfer P, Honjo S, Depetris PJ (eds) Particle flux in the ocean. SCOPE, Wiley, pp 53–69

Depetris PJ, Griffin JJ (1968) Suspended load in the Río de la Plata drainage basin. Sedimentology 11:53–60

Depetris PJ, Paolini JE (1991) Biogeochemical aspects of South American rivers: the Paraná and the Orinoco. In: Degens ET, Kempe S, Richey JE (eds) Biogeochemistry of major world rivers. SCOPE 42, Wiley, Chichester, pp 105–125

Depetris PJ, Gaiero DM, Probst JL, Hartmann J, Kempe S (2005) Biogeochemical output of rivers draining Patagonia´s Atlantic seaboard. J Coastal Res 21(4):835–844

Etchichuri MC, Remiro JR (1960) Muestras de fondo de la Plataforma Continental comprendida entre los paralelos 34° y 36° 30′ de latitud sur y los meridianos 53° 10′ y 56° 30′ de longitud oeste. Rev. Museo Argentino de Ciencias Naturales B. Rivadavia, Ciencias Geológicas 6(4):263 pp

Etchichuri MC, Remiro JR (1963) La corriente de Malvinas y los sedimentos pampeano-patagónicos. Com. Museo Argentino de Ciencias Naturales B. Rivadavia, Ciencias Geológicas I(20):11 pp

Ewing M, Lonardi A (1971) Sediment transport and distribution in the Argentine Basin. 5: Sedimentary structure of the Argentine Margin, basin and related provinces. In: Ahrens LH, Press F, Runkorn SK, Urey HC (eds) Physics and Chemistry of the Earth. Pergamon Press, London 8:125–249

Ewing M, Carpenter G, Windisch C, Ewing JI (1973) Sediment distribution in the oceans: the Atlantic. Geol Soc Am Bull 84:71–88

Franco-Fraguas P, Burone L, Mahiques MM, Ortega L, Uriene C, Muñoz A, López G, Marin Y, Carranza A, Lahuerta N, de Mello C (2014) Hydrodynamic and geomorphological controls on surface sedimentation at the Subtropical Shelf Front/Brazil-Malvinas Confluence transition off Uruguay (Southwestern Atlantic Continental Margin). Mar Geol 349:24–36

Franco-Fraguas P, Burone L, Mahiques MM, Ortega L, Carranza A (2016) Modern sedimentary dynamics in the Southwestern Atlantic Contouritic Depositional System: New insights from the Uruguayan margin based on a geochemical approach. Mar Geol 376: 15–25

Franke D, Neben S, Ladage B, Schreckenberge B, Hinz K (2007) Margin segmentation and volcano-tectonic architecture along the volcanic margin off Argentina/Uruguay, South Atlantic. Mar Geol 244:46–67

Fray Ch, Ewing M (1963) Pleistocene sedimentation and fauna of the Argentine shelf. Proc Acad Nat Sci Philadelphia 115(6):113–152

Frenz M, Höppner R, Stuut J-BW, Wagner T, Henrich R (2004) Surface sediment bulk geochemistry and grain-size composition related to the oceanic circulation along the South American continental margin in Southwest Atlantic. In: Wefer G, Mulitza S, Ratmeyer V (eds) The South Atlantic in the Late Quaternary. Springer, pp 347–373

Gaiero D, Probst J, Depetris P, Leleyter L, Kempe S (2002) Riverine transfer of heavy metals from Patagonia to the southwestern Atlantic Ocean. Reg Environ Change 3(1):51–64

Gaiero DM, Probst JL, Depetris PJ, Bidart SM, Leleyter L (2003) Iron and other transition metals in Patagopnian riverborn and windborne materials: geochemical control and transport to the South Atlantic Ocean. Geochim Cosmochim Acta 67:3603–3623

Gaiero DM, Brunet F, Probs J-L, Depetris PJ (2007) A uniform isotopic and chemical signature of dust exported from Patagonia: rock sources and occurrence in southern environments. Chem Geol 238(1–2):107–120

García Chapori N, Laprida C, Watanabe S, Totah V, Violante RA (2014) Mid-Late Pleistocene benthic foraminifera from Southwestern South Atlantic: driven by primary productivity or watermass properties? Micropaleontology 60(2):195–210
García Chapori N, Chiessi CM, Bickert T, Laprida C (2015) Sea-surface temperature reconstruction of the Quaternary western South Atlantic: new planktonic foraminiferal correlation function. Palaeogeogr Palaeoclimatol Palaeoecol 425:67–75
Garming JFL, Bleil U, Riedinger N (2005) Alteration of magnetic mineralogy at the sulfate–methane transition: analysis of sediments from the Argentine continental slope. Phys Earth Planet Inter 151:290–308
Gelós EM, Spagnuolo JO, Lizasoain GO (1988) Mineralogía y caracterización grannulométrica de sedimentos actuales de la Plataforma Argentina entre los paralelos 39° y 43° de Latitud Sur y del Golfo San Matías. Rev Asoc Geol Argentina 43(1):63–79
GeoMapApp (2013) Marine Geoscience Data System at Lamont-Doherty Earth Observatory of Columbia University. http://www.geomapapp.org
Giberto DA, Bremec CS, Acha EM, Mianzan H (2004) Large-scale spatial patterns of benthic assemblages in the SW Atlantic: the Río de la Plata estuary and adjacent shelf waters. Estuar Coast Shelf Sci 61:1–13
Gombos AM, Ciesielski PF (1977) Late Eocene to early Miocene diatoms from the southwest Atlantic. Init Rep Deep Sea Drilling Proj 71:583–634, Washington (U.S. Government Printing Office)
Gómez E, Martínez D, Cusminsky G, Suárez M, Vilanova I, Guerstien R (1992) Estudio del testigo PS2, Cuaternario del estuario de bahía Blanca, provincia de Buenos Aires. Parte I: sedimentología y micropaleontología. III Jornadas Geológicas Bonaerenses. Actas: 37–52
Gómez E, Martínez D, Cuminsky G, Borel G, Guerstein R (2000) Submerged tidal flats at the Bahía Blanca Estuary, Argentina. 17 Latein Amerika-Kolloquim Oktober (Sttugart). Profil, Band 18
Gómez E, Martínez D, Borel M, Guerstein R, Cusminsky G (2005) Submarine evidences of Holocene sea-level fluctuations in the Bahía Blanca Estuary, Argentina. J S Am Earth Sci 20:139–155
Gómez EA, Martínez DE, Borel CM, Guerstein R, Cusminsky GC (2006) Negative Sea Level Oscillation in Bahía Blanca Estuary related to a Global Climatic Change around 2650 yr B. P. J Coastal Res 39:181–185
Govin A, Holzwarth U, Heslop D, Ford Keeling L, Zabel M, Mulitza S, Collins JA, Chiessi CM (2012) Distribution of major elements in Atlantic surface sediments (36°N–49°S): imprint of terrigenous input and continental weathering. Geochem Geophys Geosyst 13:Q01013
Griffin JJ, Windom H, Goldberg ED (1968) The distribution of clay minerals in the world ocean. Deep-Sea Res 15:433–459
Groeneveld J, Chiessi CM (2011) Mg/Ca of *Globorotalia inflata* as a recorder of permanent thermocline temperatures in the South Atlantic. Paleoceanography 26(2):1–12
Groot JJ, Groot CR, Ewing M, Burckle L, Conolly JR (1965) Spores, pollen, diatoms and provenance of the Agentine basin sediments. Prog Oceanogr 4:179–217
Gruetzner J, Uenzelmann-Neben G, Franke D (2011) Variations in bottom water activity at the southern Argentine margin: indications from a seismic analysis of a continental slope terrace. Geo-Mar Lett 31:405–417
Gruetzner J, Uenzelmann-Neben G, Franke D (2012) Variations in sediment transport at the Argentina continental margin during the Cenozoic. G3 Geochem Geophys Geosyst 13(10):1–15. Q10003, doi:10.1029/2012GC004266
Gruetzner J, Uenzelmann-Neben G, Franke D (2016) Evolution of the northern Argentine margin during the Cenozoic controlled by bottom current dynamics and gravitational processes. Geochem Geophys Geosyst. doi:10.1002/2015GC006232
Guerstein GR, Vilanova FR, Suarez MI, Cusminsky GC, Martinez D, Gomez EA (1992) Estudio del testigo PS2, Cuaternario del Estuario de Bahía Blanca, provincia de Buenos Aires, Argentina. Parte II: Evaluación Paleoambiental. III Jornadas Geológicas Bonaerenses. Actas: 37–52

Guilderson TP, Burkle L, Hemming S, Peltier WR (2000) Late Pleistocene sea level variations derived from the Argentine Shelf. Geochem Geophys Geosyst 1:2000GC000098

Harloff J, Mackensen A (1996) Recent benthic foraminiferal association and ecology of the Scotia Sea and Argentain Basin. Mar Micropaleontol 31:1–29

Harris W (1977) Palynology of cores from Deep Sea Drilling Sites 327, 328, and 330, South Atlantic Ocean. Init Rep Deep Sea Drilling Proj 36:761–815

Henke S, Strasser M, Schwenk T, Hanebuth TJJ, Hüsener J, Arnold GL, Winkelmann D, Formolo M, Tomasini J, Krastel S, Kasten S (2011) An interdisciplinary investigation of a recent submarine mass transport deposit at the continental margin off Uruguay. G3 Geochem Geophys Geosyst 12:Q08009

Hernández-Molina FJ, Paterlini CM, Violante RA, Marshall P, de Isasi M, Somoza L, Rebesco M (2009) A contourite depositional system on the Argentine Slope: an exceptional record of the influence of Antarctic water masses. Geology 37(6):507–510

Hernández-Molina FJ, Paterlini CM, Somoza L, Violante RA, Arecco MA, de Isasi M, Rebesco M, Uenzelmann-Neben G, Neben S, Marshall P (2010) Giant mounded drifts in the Argentina Continental Margin: origins, and global implications for the history of thermohaline circulation. Mar Pet Geol 27:1508–1530

Hernández-Molina FJ, Soto M, Piola AR, Tomasini J, Preu B, Thompson P, Badalini G, Creaser A, Violante RA, Morales E, Paterlini CM, de Santa Ana H (2015) A contourite depositional system along the Uruguayan continental margin: sedimentary, oceanographic and paleoceanographic implications. Mar Geol. doi:10.1016/j.margeo.2015.10.008

Hinz K, Neben S, Schreckenberger B, Roeser HA, Block M, Gonzalvez de Souza K, Meyer H (1999) The Argentina continental margin north of 48°S: sedimentary successions, volcanic activity during breakup. Mar Pet Geol 16:1–25

Hodell DA, Kennett JP, Leonard KA (1983) Climatically induced changes in vertical watermass structure of the Vema Channel during the Pliocene: evidence from DSDP Sites 516A, 517 and 518. In Barker PF et al (eds) Initial Reports DSDP 72:907–919, U.S. Government Printing Office, Washington

Johnson DA (1983) Paleocirculation of the southwestern Atlantic. In Barker PF et al (eds) Initial Reports DSDP 72 907–919:977–994, U.S. Government Printing Office, Washington

Klauss A, Ledbetter MT (1988) Deep-sea sedimentary processes in the Argentine Basin revealed by high-resolution seismic records (3.5 kHz echograms). Deep Sea Res Part A Oceanogr Res Pap 35(6):899–917

Krasheninnikov VA, Basov IA (1983) Cenozoic planktonic foraminifers of the Falkland Palteau and Argentine basin, Deep Sea Drilling Project Leg 71. Init Rep Deep Sea Drilling Proj 71:821–858, Washington (U.S. Government Printing Office)

Krastel S, Wefer G, Hanebuth TJJ, Antobreh AA, Freudenthal T, Preu B, Schwenk T, Strasser M, Violante RA, Winkelmann D, M78-3 shipboard scientific party (2011) Sediment dynamics and geohazards offshore Uruguay and northern Argentina: first results from the multi-disciplinary Meteor-cruise M78-3. GeoMar Lett 31:271–283

Laprida C, Compagnucci RH, Garcia Chapori N, Violante RA (2007) Holocene evolution and paleoenvironments of the shoreface-offshore transition, northeastern Argentina: new evidences based on benthic microfauna. Mar Geol 40:43–56

Laprida C, García Chapori N, Chiessi CM, Violante RA, Watanabe S, Totah V (2011) Middle Pleistocene sea surface temperature in the Brazil-Malvinas Confluence Zone: paleoceanographic implications based on planktonic foraminifera. Micropaleontology 57(2):183–194

Laprida C, García Chapori N, Ramón Mercau J, Watanabe S, Totah V, Marcolini S, Violante RA (2014) Condicionantes ambientales de comunidades de foraminíferos bentónicos durante el Pleistoceno Medio en el Atlántico Sudoccidental. Revista de la Asociación Geológica Argentina 71(3):356–368

Lastras G, Acosta J, Muñoz A, Canals M (2011) Submarine canyon formation and evolution in the Argentina Continental Margin between 44° 30′S and 48°S. Geomorphology 128:116–136

Ledbetter MT (1986) Bottom-current pathways in the Argentine basin revealed by mean silt particle size. Nature 321:423–425

Ledbetter MT (1993) Late Pleistocene to Holocene fluctuations in bottom current-speed in the Argentine basin mudwave. Deep-Sea Res 40:911–920

Lohmann GP (1978) Abyssal benthonic foraminifera as hydrographic indicators in the western South Atlantic ocean. J Foramin Res 8:6–34

Lonardi A, Ewing M (1971) Sediment transport and distribution in the Argentine Basin. In: Press F, Runkorn SK, Urey HC (eds) Physics and Chemistry of the Earth. Pergamon Press, London 8:253–264

Mahiques MM, Tassinari CCG, Marcolini S, Violante RA, Lopes Figueira RC, Almeida da Silveira IC, Burone L, de Mello e Souza SH (2008) Nd and Pb isotope signatures on the Southeastern South America upper margin: implicances for sediment transport and source rocks. Mar Geol 250:51–63

Malumián N (1970) Bioestratigrafía del Terciario marino del subsuelo de la provincia de Buenos Aires (Argentina). Ameghiniana 7:173–204

Malumián N, Náñez C (1991) Paleogeografía del Terciario medio del cono sur: avance de aguas antarticas. In: Congreso Geológico Chileno, No. 6, Resúmenes expandidos: 847–851. Santiago de Chile

Malumián N, Nañez C (1996) Microfósiles y nanofósiles calcáreos de la plataforma continental. In: Ramos VA, Turic M (eds) Geología y Recursos Naturales de la Plataforma Continental Argentina. XIII Congreso Geológico Argentino y III Congreso de Exploración de Hidrocarburos, Buenos Aires. Asociación Geológica Argentina-Instituto Argentino del Petróleo, Relatorio 5:73–93

Mollenhauer G, McManus JF, Benthien A, Müller PJ, Eglinton TI (2006) Rapid lateral particle transport in the Argentine Basin: Molecular ^{14}C and $^{230}Th_{xs}$ evidence. Deep-Sea Res I 53:1224–1243

Morard R, Quillévéré F, Douady CJ, de Vargas C, de Garidel-Thoron T, Escarguel G (2011) Worldwide genotyping in the planktonic foraminifer *Globoconella inflata*: implications for life history and paleoceanography. PLoS ONE 6(10):26665. doi:10.1371/journal.pone.0026665

Morard R, Darling KF, Mahé F, Audic S, Ujii ., Weiner AKM, Aurore A, Seears HA, Wade CM, Quillévéré F, Douady CJ, Escarguel G, de Gardiel-Thoron T, Siccha M, Kucera M, de Vargas C (2015) PFR2: a curated database of planktonic foraminifera 18S ribosomal DNA as a resource for studies of plankton ecology, biogeography and evolution. Mol Ecol Res 15:1472–1485. doi:10.1111/1755-0998.12410

Muñoz A, Cristobo J, Rios P, Druet M, Polonio V, Uchupi E, Acosta J, Atlantis Group (2012) Sediment drifts and cold-water coral reefs in the Patagonian upper and middle continental slope. Mar Pet Geol 36(1):70–82

Nagai RH, Ferreira PAL, Mulkherjee S, Martins VM, Figueira RCL, Sousa SHM, Mahiques MM (2013) Hydrodynamic controls on the distribution of surface sediments from the southeast South American continental shelf between 23°S and 38°S. Cont Shelf Res 89:51–60

Noble TL, Piotrowski AM, Robinson LF, McManus JF, Hillenbrand C-D, Bory AJ-M (2012) Greater supply of Patagonian-sourced detritus and transport by the ACC to the Atlantic sector of the Southern Ocean during the last glacial period. Earth Planet Sci Lett 317–318:374–385

Ortiz Jauguerizar E, Cladera GA (2006) Paleoenvironmental evolution of southern South America during the Cenozoic. J Arid Environ 66:498–532

Parker G, Violante RA (1982) Geología del frente de costa y plataforma interior entre Pinamar y Mar de Ajó, Prov. de Buenos Aires. Acta Oceanográphica Argentina 3:57–91

Parker G, Perillo GME, Violante RA (1978) Características geológicas de los Bancos Alineados (Linear Shoals) frente a Punta Médanos, Prov. de Buenos Aires. Acta Oceanográphica Argentina 2:11–50

Parker G, Violante RA, Paterlini CM, Marcolini S, Costa IP, Cavallotto JL (2008) Las secuencias sismoestratigráficas del Plioceno-Cuaternario en la Plataforma Submarina adyacente al litoral del este bonaerense. Lat Am J Sedimentol Basin Anal 15:105–124

Pérez Panera JP (2012) Nanofósiles calcáreos y bioestratigrafía del Cretácico del sudeste de la cuenca Austral, Patagonia, Argentina. Ameghiniana 49(2):137–163

Pérez Panera JP (2013) Paleogene calcareous nannofossil biostratigraphy for two boreholes in the eastern Austral Basin, Patagonia, Argentina. Andean Geol 40(1):117–140

Pérez Panera JP (2015) A silicoflagellate from the middle Albian of Austral Basin, Argentina. Andean Geol 42(3):397–402. doi:10.5027/andgeoV42n3-a07

Pérez Panera JP, Angelozzi GN (2006) Nanofósiles calcáreos del Cretácico tardío–Terciario, del Pozo BB III A x-1 (Bahía Blanca), Cuenca del Colorado, Argentina. Ameghiniana 43:557–565

Pérez Panera JP, Angelozzi GN, Ronchi DI (2006) Nanofósiles calcáreos y foraminíferos del intervalo Cretácico–Paleógeno del pozo Bahía Blanca III, Cuenca del Colorado, Argentina. 9° Congreso Argentino de Paleontología y Bioestratigrafía, Córdoba. Resúmenes: 233

Pierce JW, Siegel FR (1979) Suspended particulate matter on the southern Argentina shelf. Mar Geol 29:73–91

Pivel MAG, Santarosa ACA, Toledo FAL, Costa KB (2013) The Holocene onset in the southwestern South Atlantic. Palaeogeogr Palaeoclimatol Palaeoecol 374:164–172

Potter PE (1994) Modern sands of South America: composition, provenance and global significance. Geol Rundsch 83:212–232

Preu B, Schwenk T, Hernández-Molina FJ, Violante RA, Paterlini CM, Krastel S, Tomasini J, Spieß V (2012) Sedimentary growth pattern on the northern Argentine slope: the impact of North Atlantic Deep Water on southern hemisphere slope architecture. Mar Geol 329–331:113–125

Preu B, Hernández-Molina FJ, Violante RA, Piola AR, Paterlini CM, Schwenk T, Voigt I, Krastel S, Spieß V (2013) Morphosedimentary and hydrographic features of the northern Argentine margin: the interplay between erosive, depositional and gravitational processes and its conceptual implications. Deep-Sea Res 75:157–174

Quillévéré F, Morard R, Escarguel G, Douady CJ, Ujiié Y, de Garidel-Thoron T, de Vargas C (2011) Global scale same-specimen morpho-genetic analysis of Truncorotalia truncatulinoides: a perspective on the morphological species concept in planktonic foraminifera. Palaeogeogr Palaeoclimatol Palaeoecol 391(A):2–12. doi:10.1016/j.palaeo.2011.03.013

Razik S, Govin A, Chiessi CM, von Dobeneck T (2015) Depositional provinces, dispersal, and origin of terrigenous sediments along the SE South American continental margin. Mar Geol 363:261–272

Richards HG (1966) Pleistocene mollusks from cores taken from the continental shelf of Argentina and Chile. Quaternaria 8:253–258

Richards HC, Craig JR (1963) Pleistocene mollusks from the continental shelf of Argentina. Proc Acad Nat Sci Philadelphia 115(6):127–152

Riedinger N, Pfeifer K, Kasten S, Garming JFL, Vogt C, Hensen C (2005) Diagenetic alteration of magnetic signals by anaerobic oxidation of methane related to a change in sedimentation rate. Geochim Cosmochim Acta 69(16):4117–4126

Riedinger N, Formolo MJ, Lyons TW, Henkel S, Beck A, Kasten S (2014) An inorganic geochemical argument for coupled anaerobic oxidation of methane and iron reduction in marine sediments. Geobiology 12:172–181

Schnitker D (1980) Quaternary deep-sea benthic foraminifers and bottom water masses. Annu Rev Earth Planet Sci 8:343–370

Teruggi ME (1954) El material volcánico-piroclástico en las sedimentación cuaternaria argentina. Rev Asoc Geol Argentina 9(3):184–191

Tjalsma RC (1977) Cenozoic foraminifera from the south Atlantic, DSDP Leg 36. Init Rep Deep Sea Drilling Proj 36:493–517

Toledo FA, Costa KB, Pivel MA (2007) Salinity changes in the western tropical South Atlantic during the last 30 kyr. Global Planet Change 57(3):383–395

Toledo FA, Costa KB, Pivel MA, Campos EJ (2008) Tracing past circulation changes in the western South Atlantic based on planktonic foraminifera. Rev Bras Paleontol 11(3):169–178

Urien CM, Ewing M (1974) Recent sediments and environments of southern Brazil, Uruguay, Buenos Aires and Río Negro continental shelf. In: Burk C, Drake Ch (eds) The geology of continental margins. Springer, New York, pp 157–177

Urien CM, Zambrano JJ (1996) Estructura de la Plataforma Continental. In: Ramos VA, Turic. M (eds) Geología y Recursos Naturales de la Plataforma Continental Argentina. XIII Congreso Geológico Argentino y III Congreso de Exploración de Hidrocarburos, Buenos Aires 1996. Asociación Geológica Argentina-Instituto Argentino del Petróleo, Relatorio 1:29–66

Vilanova I, Guerstein GR, Akselman R, Prieto A (2008) Mid- to Late Holocene organic walled dinoflagellate cysts from the northern Argentine shelf. Rev Palaeobot Palynol 152:11–20

Violante RA (2003) Geomorphology and coastal evolution during the last interglacial maximum sea-level highstand in the Atlantic littoral of the Pampean region, Argentina. "Puglia 2003" Final Conference Project IGCP 437 "Coastal Environmental Change during Sea-Level Highstands", Otranto-Taranto, Italia. Abstracts: 243–245

Violante RA, Parker G (1993) Evidencias de una ingresión marina intercalada entre el Interensenadense y el Belgranense ("Pampeano") en las áreas costeras y marinas del este bonaerense. XII Congre-so Geológico Argentino, Mendoza. Actas II:252–259

Violante RA, Parker G (2000) El Holoceno en las regiones costeras y marinas del noreste de la Provincia de Buenos Aires. Rev Asoc Geol Argentina 55(4):337–351

Violante RA, Parker G (2004) The post-last glacial maximum transgression in the de la Plata river and adjacent inner continental shelf, Argentina. Quat Int 114(1):167–181

Violante RA, Paterlini CM, Costa IP, Hernández-Molina J, Segovia LM, Cavallotto JL, Marcolini S, Bozzano G, Laprida C, García Chapori N, Bickert T, Spiess V (2010) Sismoestratigrafía y evolución geomorfológica del talud continental adyacente al litoral del este bonaerense, Argentina. Lat Am J Sedimentol Basin Anal 17:33–62

Violante RA, Costa IP, Cavallotto JL, Paterlini CM, Marcolini S, Bozzano G (2014a). Rasgos Morfosedimentarios, Procesos y Evolución de la Plataforma Continental Argentina desde el Último Máximo Glacial. Revista de la Asociación Geológica Argentina 71(2):292–310

Violante RA, Laprida C, Bressan GS, Diaz GR, Bozzano G, Grant JP, Garcia Chapori N, Cavallotto JL, Maidana NI, Cianfagna FA, Osterrieth ML, Paterlini CM, Costa IP (2014b) Registro paleoambiental del Estadio Isotópico 2 en la Plataforma Continental exterior del Sureste Bonaerense: un nuevo aporte a la evolución regional. Revista de la Asociación Geológica Argentina 71(3):449–455

Violante RA, Marcolini S, Cavallotto JL, Paterlini CM, Costa IP, Laprida C, Dragan, W, Watanabe S, Totah V, Rovere EI, Osterrieth ML (2014c) The Argentina continental shelf: morphology, sediments, processes and evolution since the Last Glacial Maximum. In: Chivas AR, Chiocci FL (eds) Continental Shelves during Last Glacioeustatic Cycle: Shelves of the World Reviews. Final volumen Project IUGG-UNESCO-IGCP 464 "Continental Shelves during the Last Glacial Cycle". Geological Society of London, Memoirs 41:55–68

Voigt I, Henrich R, Preu BM, Piola AR, Hanebuth TJJ, Schwenk T, Chiessi CM (2013) A submarine canyon as a climate archive—interaction of the Antarctic Intermediate Water with the Mar del Plata Canyon (Southwest Atlantic). Mar Geol 341:46–57

Voigt I, Chiessi CM, Prange M, Mulitza S, Groeneveld J, Varma V, Henrich R (2015) Holocene shifts of the southern westerlies across the South Atlantic. Paleoceanography 30(2):39–51

Voigt I, Chiessi CM, Piola AR, Henrich, R (2016). Holocene changes in Antarctic intermediate water flow strength in the Southwest Atlantic. Paleogeogr Paleoclim Paleoecol. doi:10.2016/j.palaeo.2016.09.018

von Lom-Keil H, Spieß V, Hopfauf V (2002) Fine-grained sediment waves on the western flank of the Zapiola Drift, Argentine Basin: evidence for variations in Late Quaternary bottom flow activity. Mar Geol 192:239–258

Wind FH, Wise SW (1983) Correlation of Upper Campanian-Lower Maastrichtian calcareous nannofossil assemblages in drill and piston cores from the Falkland Plateau, Southwest Atlantic Ocean. Init Rep Deep Sea Drilling Proj 71:551–563

Wise SW (1983) Mesozoic and Cenozoic calcareous nannofossils recovered by DSDP Leg 71 in the Falkland Plateau region, Southwest Atlantic Ocean. Init Rep Deep Sea Drilling Proj 71:481–550

Chapter 8
Conclusions

Abstract This chapter stresses the importance of the Argentine Margin as a potential source for paleoceanographic and paleoclimatic reconstructions in the Southern Ocean, and sets the basis of the knowledge needed for future research in the region.

Keywords Argentina Continental Margin · Archive for paleoreconstructions · Source-to-Sink · Terrigenous sedimentation · Water-masses · Sediments-oceanography interaction

The enormous extension of the Argentina Continental Margin and the great variety of environments developed on it, which reflect the many diverse characteristics that depend on its location on different geotectonic, oceanographic and morphosedimentary settings, added to the fact that it is emplaced in a key region of the Earth oceanographic-climatic system, makes the region a major target for precise paleoreconstructions in the world ocean, with implicances in local, regional and global changes. However, its study is still incomplete, and therefore a compilation of the present knowledge—as it is herein presented—is necessary for the planning of future research and applied activities in the region.

The most relevant characteristics of this margin rely on its morphology, the impressive extension of the component features (shelf, slope, rise, submarine canyons) and the highly dominant terrigenous "coarse" (sand-silt) sedimentation, not only in the nearshore and shelf areas but also on the slope and rise and even beyond (abyssal plain), with yet high sedimentation rates in many parts of the margin. This aspect is mainly conditioned by the dominantly high-energy processes that favor an active sediment transport along the Source-to-Sink cycle, that comprise not only the "terrestrial" factors like fluvial, eolian and glacial processes, but also—and of great importance—the coastal erosion and the sweeping of the shelf during the Quaternary sea-level fluctuations that favored a large delivery of sediment from the continent to the deep ocean. Energetic submarine processes such as gravitational and strong alongslope currents occurring in a context of a complex oceanographic setting—the last one mentioned reflecting the influence of

R.A. Violante et al., *The Argentina Continental Margin*,
SpringerBriefs in Earth System Sciences, DOI 10.1007/978-3-319-04196-4_8

Antarctic-sourced water-masses that flow along the margin with a high competence for transporting and distributing "coarse" sediments—are responsible of the dominant sediment characteristics. In this sense, and because of the sweeping of the shelf by the sea-level back-and-forth during the glacieustatic fluctuations, the slope and rise contain better-preserved sedimentary sequences than the shelf as providers of more adequate evidences for paleoreconstructions. On the other hand, the inter-relation "sediments—oceanographic processes" reveals a strong interaction between water-masses/marine currents and sediments, what makes the margin as highly suitable for paleoceanographic studies.

Despite the relatively "high-energy" environment for the entire margin, sedimentary, biogenic, chemical and biogeochemical processes enable to leave different categories of well-preserved proxies, tracers and records at a great variety of scales (from mega to micro) which have demonstrated to be highly valuable for paleoreconstructions.

A lot of work is still needed in the region, which constitutes a very promising setting and a highly potential archive for future research in paleoceanography and paleoclimatology at local, regional and global levels.

Further Readings

Bascom W (1980) Waves and beaches. Anchor Press, p 366

Baumann K-H, Böckel B, Donner B, Gerhardt S, Henrich R, Vink A, Volbers A, Willems H, Zonneveld KAF (2004) Contribution of calcareous plankton groups to the carbonate budget of South Atlantic surface sediments. In: Wefer G, Mulitza S, Ratmeyer V (eds) The South Atlantic in the Late Quaternary. Springer, Germany, pp 81–99

Berger WH, Schulz M, Wefer G (2010) Quaternary oceans and climate change: lessons for the future? Int J Earth Sci (Geol Rundsch) 99 (1):S171–S189

Bianchi AA, Garzoli SL (1997) Variabilidad y movimientos del frente Brasil-Malvinas/Variability and motion of the Brazil-Malvinas front. Geoacta 22:74–90

Boltovskoy E (1980) On the benthonic bathyal-zone Foraminifera as stratigraphic guide

Bond GC, Kominz MA (1988) Evolution of thought on passive continental margins from the origin of geosynclinal theory ($\sim$1860) to the present. Geol Soc Am Bull 100:1909–1933

Bourcart J (1952) Océanographie. Introduction á la Géologie. CDU. L´information géographique 16(4):163

Bratzeva GM (1983) Spores and Pollen from Cenozoic sediments of the Falkland Plateau. Initial reports of the deep sea drilling project 71, Washington (U.S. Government Printing Office), pp 907–932

Calvert SE, Pedersen TF (2007) Elemental proxies for paleoclimatic and paleoceanographic variability in marine sediments: interpretation and application. In: Hillaire-Marcel C, de Vernal A (eds) Proxies in late cenozoic paleoceanography. developments in marine geology 1(14):567–644

Carvalho L, Jones C, Liebmann B (2004) The South Atlantic convergence zone: intensity, form, persistence, and relationships with intraseasonal to interannual activity and extreme rainfall. J Climate 17:88–108

Catuneanu O (2002) Sequence stratigraphy of clastic systems: concepts, merits, and pitfalls. J Afr Earth Sci 35(1):1–43

Catuneanu O (2006) Principles of sequence stratigraphy (1st ed), Elsevier, 375 pp

Christelle C, Hamelin B (2007) Isotopic tracers of water masses and deep currents. In: Hillaire-Marcel C, de Vernal A (eds) Proxies in late cenozoic paleoceanography. Develop Mar Geol 1(15): 645–680
Clapperton C (1993) Quaternary geology and geomorphology of South America. Elsevier Sciences, New York 779 pp
COMARGE (2010) Continental margins ecosystems on a worldwide scale. COMARGE–IFREMER, France. http://www.ifremer.comarge/en/index.html
Combes V, Matano RP (2014) Trends in the Brazil/Malvinas confluence region. Geophys Res Lett 41(24):8971–8977
Curray JR (1975) Marine sediments, geosynclines and orogeny. In: Fischer A, Judson S (eds) Petroleoum and global tectonics. Princeton University Press, New Jersey, pp 157–122
DeMaster DJ (2002) The accumulation and cycling of biogenic silica in the Southern Ocean: revisiting the marine silica budget. Deep-Sea Res II 49:3155–3167
Dorschel B, Gutt J, Piepenburg D, Schröder M, Arndt JE (2014) The influence of the geomorphological and sedimentological settings on the distribution of epibenthic assemblages on a flat topped hill on the over-deepened shelf of the western Weddell Sea (Southern Ocean). Biogeosciences 11:3797–3817
Dunlop P, Sacchetti F, Benetti S, O'Cofaigh C (2011) Mapping Ireland's glaciated continental margin using marine geophysical data. In: Smith MJ, Paron P, Griffiths JS (eds) Geomorphological mapping. Develop Earth Surf Process 15:339–354
Duxbury AC (1971) The earth and its oceans. Addison Wesley, USA 381 pp
Fetter AF, Matano RP (2008) On the origins of the variability of the Malvinas current in a global, eddy-permitting numerical simulation. J Geophys Res Oceans (1978–2012) 113 (C11)
Figueroa SN, Satyamurty P, da Silva Leite, Dias P (1995) Simulations of the summer circulation over the South American Region with the ETA coordinate model. Am Meteorol Soc 52: 1573–1584
Flantua SGA, Hooghiemstra H, Vuille M, Behling H, Carson JF, Gosling WD, Hoyos I, Ledru MP, Montoya E, Mayle F, Maldonado A, Rull V, Tonello MS, Whitney BS, Gonzalez-Arango C (2015) Climate variability and human impact on the environment in South America during the last 2000 years: synthesis and perspectives. Climate Past Discussions 11:3475–3565. doi:10.5194/cpd-11-3475-2015
Francois R (2007) Paleoflux and paleocirculation from sediment ^{230}Th and ^{231}Pa/^{230}Th. In: Hillaire-Marcel C, de Vernal A (eds) Proxies in Late cenozoic p. Develop Mar Geol 1(16): 681–716
Fu R, Zhu B, Dickinson RE (1998) How do atmosphere and land surface influence seasonal changes of convection in the tropical Amazon? J Climate 12:1306–1321
Gappa JL (2000) Species richness of marine Bryozoa in the continental shelf and slope off Argentina (south-west Atlantic). Diversity and distributions 6(1):15–27
García Chapori N (2015) Reconstrucción paleoceanográfica del talud bonaerense a partir de testigos cuaternarios. Ph.D. thesis, University of Buenos Aires, p 203
Garreaud RD, Vuille M, Compagnucci R, Marengo J (2009) Present-day South American climate. Palaeogeogr Palaeocl 281:180–195. doi:10.1016/j.palaeo.2007.10.032
Garrison T (2008) Essentials of oceanography. Brooks/Cole Belmont, CA, USA. ISBN-13: 978-0-8400-6155-3 ISBN-10: 0-8400-6155-2
Garzoli SL, Giulivi C (1994) What forces the variability of the southwestern Atlantic boundary currents? Deep Sea Res Part I: Oceanograph Res Papers 41(10):1527–1550
Garzoli SL, Matano R (2011) The South Atlantic and the Atlantic Meridional overturning circulation. Deep-Sea Res Part II 58:1837–1847
Goni GJ, Wainer I (2001) Investigation of the Brazil current front variability from altimeter data. J Geophys Res: Oceans (1978–2012) 106(C12):31117–31128
Goni GJ, Bringas F, DiNezio PN (2011) Observed low frequency variability of the Brazil Current front. J Geophys Res: Oceans (1978–2012) 116(C10)
Gordon AL (1981) South Atlantic thermocline ventilation. Deep Sea Res Part A. Oceanograp Res Papers 28(11):1239–1264

Gordon AL (1989) Brazil-Malvinas confluence-1984. Deep-Sea Res 36:359–384
Goudie AS (ed) (2004) Encyclopedia of geomorphology. International Assoc. of Geomorphologists, Routledge, 1156 pp
Hartnett HE, Keil RG, Hedges JI, Devol AH (1998) Influence of oxygen exposure time on organic carbon preservation in continental margin sedimentation. Nature 391:572–574
Hay WH (2008) Evolving ideas about the cretaceous climate and ocean circulation. Cretac Res 29:725–753
Heezen BC, Menard HW (1996) Topography of the deep-sea floor. In: Hill MN (ed) The sea. Intersciences Publ, 2nd Edn, J Wiley 233–280
Higginson MJ (2009) Geochemical proxies (non-Isotopic). In: Gornitz V (ed) encyclopedia of paleoclimatology and ancient environments. Part of the series Encyclopedia of Earth Sciences Series. Springer, Germany, 341–354
Hillaire-Marcel C, de Vernal A (2007) Proxies in late cenozoic paleoceanography. Develop Mar Geol 1:843
Horel J, Hahmann A, Geisler J (1989) An investigation of the annual cycle of the convective activity over the tropical Americas. J Climate 2:1388–1403
Jeffreys H (1925) On the formation of water waves by wind. Proceedings of the Royal Society of London 107A:189–206
Johnson DW (1919) Shore processes and shoreline development. Wiley, New York
Kodama Y (1992) Large-scale common features of subtropical precipitation zones (the Baiu Frontal Zone, the SPCZ, and the SACZ) part I: characteristics of subtropical frontal zones. Journal of the Meteorological Society of Japan 7:813–836
Komar PD (1976) Beach processes and sedimentation. Prentice Hall Inc, New Jersey, p 429
Ledbetter MT (1986) Bottom-current pathways in the Argentine basin revealed by mean silt particle size. Nature 321:423–425
Ledbetter MT (1993) Late pleistocene to holocene fluctuations in bottom current-speed in the Argentine basin mudwave. Deep-sea Research 40:911–920
Legeckis R, Gordon AL (1982) Satellite observations of the Brazil and Falkland currents—1975, 1976 and 1978. Deep Sea Res Part A. Oceanograp Res Papers 29(3):375–401
Liebmann B, Kiladis G, Marengo J, Ambrizzi T, Glick JD (1999) Submonthly convective variability over South America and the South Atlantic Convergence Zone. J Climate 11: 2898–2909
Lidz BH, Shinn EA (1991) Paleoshorelines, reefs, and a rising sea: South Florida, U.S.A. J Coast Res 7:203–229
Liu KK, Atkinson L, Quiñones R, Talaue McManus L (2010) Biogeochemistry of continental margins in a global context. In: Liu KK, Atkinson L, Quiñones R, Talaue McManus L (eds) Carbon and nutrient fluxes in continental margins: a global synthesis. The IGBP Series, Springer, Germany, 3–24
Lutjeharms JRE, Meeuwis JM (1987) The extent and variability of South-East Atlantic upwelling. S Afr J Mar Sci 5(1):51–62
Lutjeharms JRE, Valentine HR (1987) Water types and volumetric considerations of the South-East Atlantic upwelling regime. S Afr J Mar Sci 5(1):63–71
Lyons R, Tooth S, Duller GAT (2014) Late Quaternary climatic changes revealed by luminescence dating, mineral magnetism and diffuse reflectance spectroscopy of river terrace palaeosols: a new form of geoproxy data for the southern African interior. Quater Sci Rev 95:43–59
Marengo J, Liebmann B, Kousky VE, Filizola N, Wainer IC (2001) Onset and end of the rainy season in the Brazilian Amazon basin. J Climate 14:833–852
MARGINS (2000) Source-to-sink studies. Margins-NSF Program, Newsletter 5. http://www.nsf-margins.org
Matano RP (1993) On the separation of the Brazil current from the coast. J Phys Oceanograp 23(1):79–90
Matano RP, Palma ED (2008) On the upwelling of downwelling currents. J Phys Oceanograph 38(11):2482–2500

Matano RP, Schlax MG, Chelton DB (1993) Seasonal variability in the southwestern Atlantic. J Geophys Res: Oceans (1978–2012) 98(C10):18027–18035

McAdoo BG, Pratson LF, Orange DL (2000) Submarine landslide geomorphology, US continental slope. Mar Geol 169:103–136

McCave IN (2002) Sedimentary settings on continental margins: an overview. In: Wefer G, Billet D, Hebeln D, Jorgensen BB, Schluter M, van Weering TCE. Ocean Margin Syst, 1–14

McCave IN (2007) Deep-sea sediment deposits and properties controlled by currents. In: Hillaire-Marcel C, de Vernal A (eds) Proxies in late cenozoic paleoceanography. Develop Mar Geol 1(1):19–62

McCave IN, Manighetti B, Beveridge NAS (1995) Circulation in the glacial North Atlantic inferred from grain-size measurements. Nature 374:149–152

Meno L, Sibuet M, Carney RS, Levin L, Rowe GT, Billet DSM, Poore G, Kitazato H, Vanreusel A, Geléron J, Lavrado HP, Sellanes J, Ingole B, Krylova E (2010) New perceptions of continental margins biodiversity. In: McIntyre A (ed) Life in the World Ocean: diversity, distribution and abundance. Willey Blackwell, 384 pp

Micallef A, Berndt C, Mason DG, Stow DAV (2007) Fractal Statistics of the storega slide. In: Lykousis V, Sakellariou D, Locat J (eds) Submarine mass movements and their consequences. Springer, The Nederlands, pp 3–10

Mitchell AH, Reading HG (1969) Continental margins, geosynclines and ocean floor spreading. J Geol 77:629–646

Möller OO, Piola AR, Freitas AC, Campos EJ (2008) The effects of river discharge and seasonal winds on the shelf off southeastern South America. Cont Shelf Res 28(13):1607–1624

Munk WH, Wunsch C (1998) Abyssal recipes II: energetics of tidal and wind mixing. Deep-Sea Res 45:1977–2010

Musard O, Le Dû-Blayo L, Francour P, Beurier J-P, Feunteun E, Talassinos L (eds) (2014) Underwater seascapes. From geographical to ecological perspectives. Springer, Science and Perspectives, Berlin, 291 pp

Nakamura H, Shimpo A (2004) Seasonal variations in the southern hemisphere storm tracks and jet streams as revealed in a reanalysis dataset. J Climate 17:1828–1844

Neumann G, Pierson W Jr (1966) Principles of physical oceanography. Prentice-Hall Inc, N.J., 545 pp

Nittrouer CA, Austin Jr JA, Field ME, Kravits JH, Syvitski JPM, Wiberg P (2007) Writing the Roseta Stone: insights into continental-margin sedimentary processes and strata. In: Nittrouer CA, Austin Jr JA, Field ME, Kravits JH, Syvitski JPM, Wiberg P (eds) Continental margin sedimentation, from sediment transport to sequence stratigraphy. International Association of Sedimentologists, Sp. Publ. N 37, Blackwell Publishing, pp 1–48

Olson DB, Podestá GP, Evans RH, Brown OB (1988) Temporal variations in the separation of Brazil and Malvinas currents. Deep Sea Res Part A. Oceanograph Res Papers 35(12):1971–1990

Palma ED, Matano RP, Piola AR (2008). A numerical study of the Southwestern Atlantic shelf circulation: stratified ocean response to local and offshore forcing. J Geophys Res: Oceans (1978–2012), 113(C11)

Peterson RG, Stramma L (1991) Upper-level circulation in the South Atlantic Ocean. Progress in Oceanography 26:1–73

Pickering KT, Clark JD, Smith RDA, Hiscott RN, Ricci Lucchi F, Kenyon NH (1995) Architectural element analysis of turbidite systems, and selected topical problems for sand-prone deep-water systems. In: Pickering KT, Hiscott RN, Kenyon NH, Ricci Lucchi F, Smith RDA (eds) Atlas of deep-water environments: architectural style in turbidite systems. Chapman et al., London, pp 1–10

Pickering KT, Hiscott RN, Hein FJ (1989) Deep-marine environments, clastic sedimentation and tectonics. Unwin Hyman, London 416 pp

Piola AR, Gordon AL (1989) Intermediate waters in the southwest South Atlantic. Deep Sea Res Part A. Oceanograph Res Papers 36(1):1–16

Piola AR, Matano RP (2001) Brazil and Falklands (Malvinas) currents. Ocean currents: a derivative of the encyclopedia of ocean sciences, pp 35–43
Piola AR, Campos EJD, Möller OO, Charo M, Martínez C (2000) Subtropical shelf off eastern South America. J Geophys Res 105:6565–6578
Piola AR, Möller OO, Guerrero RA, Campos EJ (2008) Variability of the subtropical shelf front off eastern South America: Winter 2003 and summer 2004. Cont Shelf Res 28(13):1639–1648
Piola AR, Avellaneda NM, Guerrero RA, Jardón FP, Palma ED, Romero SI (2010) Malvinas-slope water intrusions on the northern Patagonia continental shelf. Ocean Sci 6(1):345
Rebesco M, Camerlenghi A (2008) Contourites. Elsevier, London 663 pp
Reboita MS, Gan MA, da Rocha RP, Ambrizzi T (2010) Regimes de precipitação na América do sul: uma revisão bibliográfica. Revista Brasileira de Meteorologia 25:185–204
Revel M, Cremer M, Grousset FE, Labeyrie L (1996) Grain-size and Sr-Nd isotopes as tracer of paleo-bottom current strength, Northeast Atlantic Ocean. Mar Geol 131:233–249
Robertson AW, Mechoso CR (2000) Interannual and interdecadal variability of the South Atlantic convergence zone. Mon Weather Rev 128:2947–2957
Sandström JW (1916) Meteorologische studien à schedischen hochgebirge. Göteborgs K. Vetensk. Vitterhets-Samh. Handl. 10:2–42
Schloesser F, Furue R, McCreary JP, Timmermann A (2012) Dynamics of the Atlantic meridional overturning circulation. Part 1: Buoyancy-forced response. Prog Oceanograph 101(1):33–62
Schmittner A, Chiang JCH, Hemming SR (2007) Introduction: the ocean's meridional overturning circulation. In Schmittner A, Chiang JCH, Hemming SR (eds) Ocean circulation: mechanisms and impacts - past and future changes of meridional overturning. American Geophysical Union, Washington, D.C., pp 1–4. doi: 10.1029/173GM02
Seibold E, Berger WH (1982) The sea floor. An introduction to marine geology. Springer, Berlin, 288 pp
Shannon LV, Agenbag JJ, Walker ND, Lutjeharms JRE (1990) A major perturbation in the Agulhas retroflection area in 1986. Deep Sea Res Part A. Oceanograph Res Papers 37(3):493–512
Shepard FP (1948) Submarine geology. Harper & Row, New York
Solman SA (2013) Regional climate modeling over South America: a review. Adv Meteorol 504357. doi:10.1155/2013/504357
St-Onge G, Mulder T, Francus P, Long B (2007) Continuous physical properties of cored marine sediments. In: Hillaire-Marcel C, de Vernal A (eds) Proxies in late cenozoic paleoceanography. Develop Mar Geol 1(2):63–98
Stow DAV, Pudsey CJ, Howe JA, Faugères J-C, Viana AR (eds) (2002) Deep-water contourite systems: modern drifts and ancient series, seismic and sedimentary characteristics. The Geological Society of London, Memoir 22, 466 pp
Stramma L, England M (1999) On the water masses and mean circulation of the South Atlantic Ocean. J Geophys Res: Oceans (1978–2012) 104(C9):20863–20883
Thomas DSG (2013) Reconstructing palaeoclimates and palaeoenvironments in drylands: what can landform analysis contribute? Earth Surf Process Landforms 38(1):3–16
Thomas DSG, Burrough SL (2012) Interpreting geoproxies of Late Quaternary climate change in African drylands: implications for understanding environmental change and early human behaviour. Quater Int 253:5–17
Toggweiler JR, Samuels BL (1993) Is the magnitude of the deep outflow from the Atlantic Ocean actually governed by Southern Hemisphere winds? In: Heimann M (ed) The global carbon cycle. Springer, Berlin, pp 303–331
Toggweiler JR, Samuels BL (1995) Effect of Drake passage on the global thermohaline circulation. Deep Sea Res Part I: Oceanograph Res Papers 42:477–500
Toggweiler JR, Samuels BB (1998) On the ocean large-scale circulation near the limit of no vertical mixing. J Phys Oceanograph 28:1832–1852
Vail PR (1987) Seismic stratigraphy interpretation procedure. In Bally AW (ed) Atlas of seismic stratigraphy: AAPG Stud Geol 27(1):1–10

Vail PR, Todd RG, Sangree JB (1977) Seismic stratigraphy and global changes of sea level: Part 5. Chronostratigraphic Significance of seismic reflections: section 2. Application of seismic reflection configuration to stratigraphic interpretation, Memoir 26:99–116

Vera C, Higgins W, Amador J, Ambrizzi T, Garreaud R, Gochis D, Gutzler D, Lettenmaier D, Marengo J, Mechoso CR, Nogues-Paegle J, Silva Diaz PL, Zhang C (2006) Towards a unified view of the American monsoon system. J Climate 19:4977–5000

Wainer I, Gent P, Goni G (2000) Annual cycle of the Brazil-Malvinas confluence region in the National Center for Atmospheric Research climate system model. J Geophys Res: Oceans (1978–2012) 105(C11):26167–26177

Wefer G, Berger WH (1996) Global change and marine geology: introduction. Geol. Rundsch. 85:399–400

Wefer G, Billet D, Hebeln D, Jorgensen BB, Schluter M, van Weering TCE (2002) Ocean margin systems. Springer, Berling-Heidelberg 495 pp

Wollast R (2002) Continental margins: a review of geochemical settings. In: Wefer G, Billet D, Hebeln D, Jorgensen BB, Schluter M, van Weering TjCE (eds) Ocean margin systems, pp 15–31

You Y (2002) Quantitative estimate of Antarctic intermediate water contributions from the Drake passage and the southwest Indian Ocean to the South Atlantic. J Geophys Res: Oceans (1978–2012)107(C4):6–1

Zhou J, Lau KM (1998) Does a monsoon climate exist over South America? J Climate 11: 1020–1040

MIX
Papier aus verantwortungsvollen Quellen
Paper from responsible sources
FSC® C105338

If you have any concerns about our products,
you can contact us on
ProductSafety@springernature.com

In case Publisher is established outside the EU,
the EU authorized representative is:
Springer Nature Customer Service Center GmbH
Europaplatz 3, 69115 Heidelberg, Germany

Printed by Libri Plureos GmbH
in Hamburg, Germany